Adaptive High-Resolution Sensor Waveform Design for Tracking

Synthesis Lectures on Algorithms and Software in Engineering

Editor
Andreas Spanias, *Arizona State University*

Adaptive High-Resolution Sensor Waveform Design for Tracking
Ioannis Kyriakides, Darryl Morrell, and Antonia Papandreou-Suppappola
2010

MATLAB® Software for the Code Excited Linear Prediction Algorithm: The Federal Standard-1016
Karthikeyan N. Ramamurthy, Andreas S. Spanias
2010

OFDM Systems for Wireless Communications
Adarsh B. Narasimhamurthy, Mahesh K. Banavar, Cihan Tepedelenliougl u
2010

Advances in Modern Blind Signal Separation Algorithms: Theory and Applications
Kostas Kokkinakis, Philipos C. Loizou
2010

Advances in Waveform-Agile Sensing for Tracking
Sandeep Prasad Sira, Antonia Papandreou-Suppappola, Darryl Morrell
2008

Despeckle Filtering Algorithms and Software for Ultrasound Imaging
Christos P. Loizou, Constantinos S. Pattichis
2008

Adaptive High-Resolution Sensor Waveform Design for Tracking

Ioannis Kyriakides, Darryl Morrell, and Antonia Papandreou-Suppappola

ISBN: 978-3-031-00387-5 paperback
ISBN: 978-3-031-01515-1 ebook

DOI 10.1007/978-3-031-01515-1

A Publication in the Springer series
SYNTHESIS LECTURES ON ALGORITHMS AND SOFTWARE IN ENGINEERING

Lecture #4
Series Editor: Andreas Spanias, *Arizona State University*
Series ISSN
Synthesis Lectures on Algorithms and Software in Engineering
Print 1938-1727 Electronic 1938-1735

Adaptive High-Resolution Sensor Waveform Design for Tracking

Ioannis Kyriakides
University of Nicosia

Darryl Morrell
Arizona State University

Antonia Papandreou-Suppappola
Arizona State University, Polytechnic Campus

SYNTHESIS LECTURES ON ALGORITHMS AND SOFTWARE IN ENGINEERING #4

ABSTRACT

Recent innovations in modern radar for designing transmitted waveforms, coupled with new algorithms for adaptively selecting the waveform parameters at each time step, have resulted in improvements in tracking performance. Of particular interest are waveforms that can be mathematically designed to have reduced ambiguity function sidelobes as their use can lead to an increase in the target state estimation accuracy Moreover, adaptively positioning the sidelobes can reveal weak target returns by reducing interference from stronger targets.

The manuscript provides an overview of recent advances in the design of multicarrier phase-coded waveforms based on Bjorck constant-amplitude zero-autocorrelation (CAZAC) sequences for use in an adaptive waveform selection scheme for mutliple target tracking. The adaptive waveform design is formulated using sequential Monte Carlo techniques that need to be matched to the highly resolution measurements.

The work will be of interest to both practitioners and researchers in radar as well as to researchers in other applications where high resolution measurements can have significant benefits.

KEYWORDS

multiple target tracking, particle filtering, radar tracking, waveform design, adaptive waveform selection

Contents

CHAPTER 1

Introduction

Recent innovations in waveform design, coupled with adaptive waveform configuration and selection capabilities, have enhanced tracking performance in modern radar systems. The goal of waveform design is to achieve low ambiguity function sidelobes, increasing target state estimation accuracy. The adaptive configuration and selection of the transmitted radar waveform can reduce interference in a multiple tracking environment enhancing the visibility of weak targets.

1.1 RADAR TRACKING WITH WAVEFORM DESIGN AND WAVEFORM SELECTION

Measurement accuracy in target range and range rate depends on the type of the transmitted waveform and the configuration of its parameters. The more localized the ambiguity function (AF) [1] of a waveform is in the delay-Doppler space the more accurate information it provides on the range and range rate of a given target with respect to the radar system. Therefore, waveform design is essential in radar system development.

The adaptive selection of the waveform transmitted at every time step of the tracking scenario can also provide significant benefits. In fact, adaptive waveform selection may become necessary in some scenarios. For example, in the case of tracking multiple targets one important problem is to observe weak targets in the presence of significantly stronger target measurements. In this case, large measurement sidelobes from stronger targets mask the weaker target measurements. Therefore, there is a need to design configurable radar waveforms and develop an adaptive radar sensor configuration technique to position strong target sidelobes away from predicted locations of weak targets.

1.2 REVIEW OF ADVANCES IN WAVEFORM DESIGN AND ADAPTIVE WAVEFORM SELECTION FOR RADAR TARGET TRACKING

The main aspects affecting radar tracking performance are the method of processing the radar measurements, the type and configuration of the transmitted waveform, and the ability to adaptively select and configure the waveform to be transmitted. The traditional method for processing radar signal returns is to separate the delay-Doppler space into tessellating shaped resolution cells centered on a fixed grid of delay and Doppler locations. These cells are constructed in such a way as to approximate the shape of the probability of detection contour formed by thresholding the ambiguity function of the transmitted waveform [2, 3]. A detection decision is then taken for each resolution cell based on the thresholded output of a matched filter placed on the centroid of each cell. The need to approximate resolution cells as parallelograms and exhaustively evaluate cells in the entire delay-Doppler space is avoided with the use of a particle filter [4] based approach. Using the matched filter locations proposed from the particle filter and the exact shape of the probability of detection contour, the exact and relevant resolution cells are interrogated [5].

Concerning the design of the waveform to be transmitted, an extensive review of different types of waveforms and their AFs is provided in [1]. The effect on tracking performance when adjusting the parameters of LFM waveforms and combining LFM with constant frequency waveforms was investigated in [2, 3]. Recently, Björck constant amplitude zero-autocorrelation (CAZAC) [6, 7] waveforms with highly concentrated AFs were used in tracking exhibiting significant gains in tracking performance [5]. Moreover, an analysis characterizing the positioning of AF sidelobes of multi-carrier phase coded (MCPC) [1] waveforms formed by combining CAZAC waveforms was provided in [8, 9].

Instead of using a fixed waveform, adaptive waveform techniques were used to minimize either the tracking error or the validation gate volume in [10, 11]. In [12, 13], adaptive waveform techniques were developed for non-linear system models with a single target and using frequency-modulated waveforms. The result was techniques that accurately represent physical systems with non-linear characteristics.

Adaptive waveform selection was also extended to the multitarget case, and specifically in the case of tracking weak targets in the presence of strong targets. When tracking multiple targets with radar sensors, weak targets are often difficult to observe. This is because the sidelobes of the AFs of strong targets are higher than the mainlobe of the AFs of the weak targets in the delay-Doppler plane. In [8, 9] configurable radar waveforms were designed and an adaptive radar sensor configuration technique was developed to select parameters of MCPC Björck CAZACs at each time step of the tracking scenario in order to minimize the predicted tracking error.

1.3 ORGANIZATION

The work is organized as follows. In Chapter 2, we describe the waveforms used in tracking applications and explain the construction of MCPC waveforms and derive their AF. In Chapter 3, we provide an overview of tracking methods, an introduction to particle filtering for single and multiple target tracking, and basic radar tracking concepts. In Chapter 4, we describe the SIRPF and LPF algorithms used in the single target tracking case with the LFM and CAZAC sequences respectively. In Chapter 5, we describe the IPLPF algorithm and a scheme for adaptively selecting the radar waveform to minimize the predicted root-mean-squared error in a multiple target tracking scenario. In Chapter 6, we draw the general conclusions from this work. In Table 1.1, we provide a list of acronyms used in this work.

Table 1.1: List of Acronyms

Acronym	Description
AF	Ambiguity function
CAZAC	Constant amplitude zero autocorrelation
EKF	Extended Kalman filter
FT	Fourier transform
FFT	Fast Fourier transform
IP	Independent partitions
IPLPF	Independent partitions likelihood particle filter
JMPD	Joint multitarget probability density
LFM	Linear frequency-modulated
LPF	Likelihood particle filter
MC	Monte Carlo
MCPC	Multicarrier phase-coded
OFDM	Orthogonal frequency division multiplexing
PF	Particle filter
RMSE	Root mean squared error
SCPC	Single-carrier phase-coded
SIR	Sequenftial importance resampling
SIRPF	Sequential importance resampling particle filter
SIS	Sequential importance sampling
SNR	Signal-to-noise ratio
SMC	Sequential Monte Carlo
UKF	Unscented Kalman filter

CHAPTER 2

Radar Waveform Design

In this chapter, we describe LFM sequences commonly used in tracking. Moreover, we present two types of waveforms that have recently been adopted in radar tracking. The first is the family of Björck CAZACs which have many useful properties, such as highly peaked AFs, that make them a favorable choice in tracking with radar. A single target tracking application using LFM and Björck CAZAC sequences is provided in Chapter 4. We also describe the construction of MCPC Björck CAZAC waveforms and derive their AFs. An adaptive selection of MCPCs is particularly useful when tracking multiple targets of different cross-sectional areas. A multiple tracking application with adaptively selected MCPC Björck CAZAC waveforms is provided in Chapter 5.

2.1 LFM SEQUENCES

In an LFM sequence, the frequency linearly increases or decreases with the duration of the waveform. Thus, LFM signals are popular because they are easy to generate [14]. In an LFM, the frequency linearly increases with the duration of the waveform. The LFM is described by the equation

$$s(m) = \exp\left(j2\pi \frac{(f_b - f_a)((m-1)^2 - (\frac{M}{2} - 1)^2)}{2(M-1)}\right)$$

for $m = 1, \ldots, M$, where f_a and f_b are, respectively, the initial and final normalized frequencies of the waveform. The normalized frequency is equal to the actual frequency in Hz divided by the sampling frequency. Thus, the minimum and maximum normalized frequencies are 0 and 0.5, respectively.

In Figure 2.1, a plot of an LFM waveform is given in the time domain. The waveform is shown in the time-frequency domain in Figure 2.2. In Figure 2.3, we show the ambiguity function of the LFM waveform.

2.2 BJÖRCK CAZAC SEQUENCES

A CAZAC sequence $s(m)$ with finite length M, has constant magnitude, $|s(m)| = 1$, $m = 0, \ldots, M-1$, and zero-autocorrelation, $(1/M)\sum_{m=0}^{M-1} s(\tau + m)\, s^*(m) = 0$, for $\tau \neq 0$, where the addition is modulo M [16, 15]. An example of a CAZAC sequence with quadratic phase is the Björck CAZAC sequence. For prime length $M = 1, \mod 4$, it is given by [6, 16]

$$s(m) = e^{j2\pi \arccos(1/(1+\sqrt{M}))\,[(m/M)]}, \quad m = 0, \ldots, M-1, \tag{2.1}$$

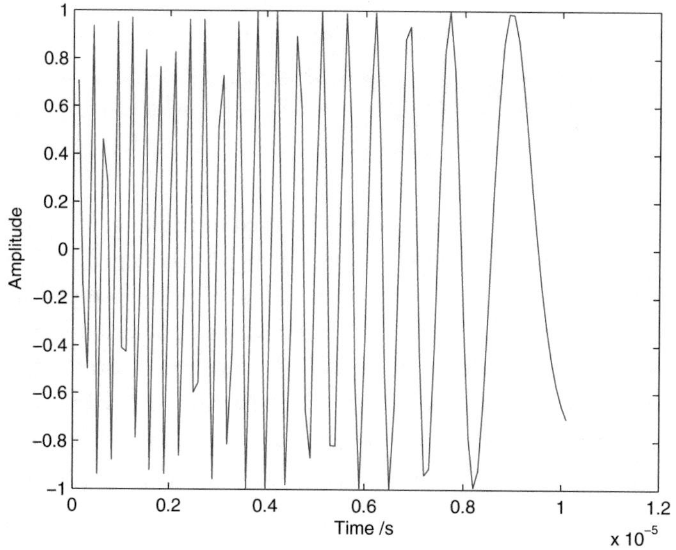

Figure 2.1: An LFM waveform with decreasing frequency along the time axis.

where m, $\bmod M$ (or m modulo M) is the remainder of the division m/M, and $[(m/M)]$ is the Legendre symbol that is given by

$$[(m/M)] = \begin{cases} 1, & \text{if } m^{(M-1)/2} = 1, \mod M \\ -1, & \text{if } m^{(M-1)/2} = -1, \mod M \\ 1, & \text{if } m = 0, \mod M \end{cases} .$$

Björck CAZAC sequences are an attractive choice for target tracking with radar [16] as their constant amplitude allows for continuous transmission of peak power and can thus lead to increases in signal-to-noise ratio (SNR). They also exhibit very tight localization in the delay-Doppler plane that can enhance the range resolution and range-rate resolution of the measurements. The discrete AF of a Björck CAZAC sequence is given by [17]

$$\mathrm{AF}_s(\tau, \nu) = \frac{1}{M} \sum_{m=0}^{M-1} s(m-\tau)\, e^{j2\pi m\, \nu/M}\, s^*(m)\,, \tag{2.2}$$

where n and ν are the discrete delay and Doppler parameters, respectively. The narrowband AF expression in (2.2) ignores the time scaling of the return signal due to the target range-rate. This assumption is valid when $WT_d \ll \frac{c}{2\dot{r}}$, where c is the propagation velocity, $\dot{r}$ is the target range rate [18, pg 241], and $W = \frac{1}{T_b}$ and $T_d = MT_b$ are the bandwidth and duration of the transmitted signal, respectively. Moreover, T_b is the bit interval. Thus, the time-bandwidth product $WT_d = M$

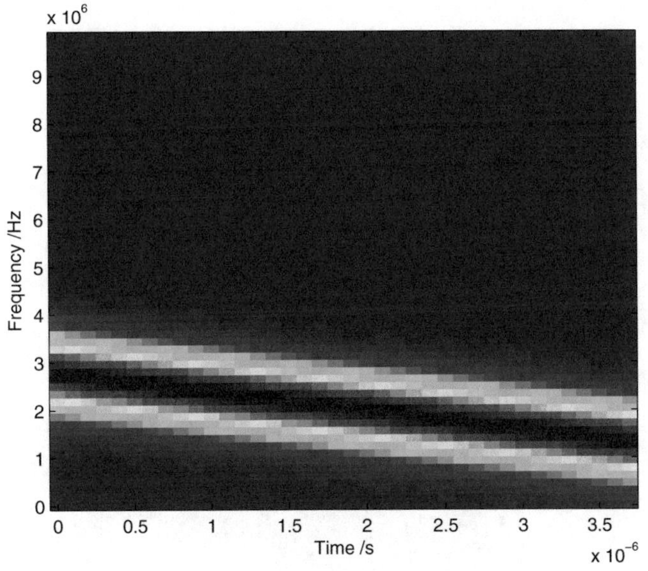

Figure 2.2: Spectrogram of an LFM waveform. This time-frequency representation shows how the frequency of this particular waveform decreases linearly with time.

is much smaller than $\frac{c}{2r}$ as the speed of propagation in the atmosphere is large. Note that phase coding is required in order to reduce the bandwidth of the CAZAC sequence to meet transmission requirements.

The AF of a exhibits a large spike at the origin $(\tau, \nu) = (0, 0)$ of the discrete delay-Doppler plane, with very small sidelobes. An example of the AF of a Björck CAZAC of length $M = 293$ is shown in Fig. 2.4.

2.3 MULTICARRIER PHASE-CODED BJÖRCK CAZAC SEQUENCES

In order to unmask weak targets in multitarget tracking scenarios, we describe waveforms whose AF can be adaptively shaped by configuring the waveform's parameters. We utilize the property that a frequency cyclically shifted CAZAC is still a CAZAC [15] with a highly concentrated AF. We show that a combination of frequency cyclically shifted sequences gives rise to AF surfaces that have sidelobes whose locations depend on the difference in cyclic shift, the number of sequences, and their length. Although cyclic permutations of CAZACs are possible in both the time and frequency domains, we choose to construct combinations of CAZACs shifted only in frequency. This is because the frequency shifts result in wide zero regions in the AF plane and facilitate the adaptive positioning of the AF sidelobes. In this section, we describe the construction of MCPC

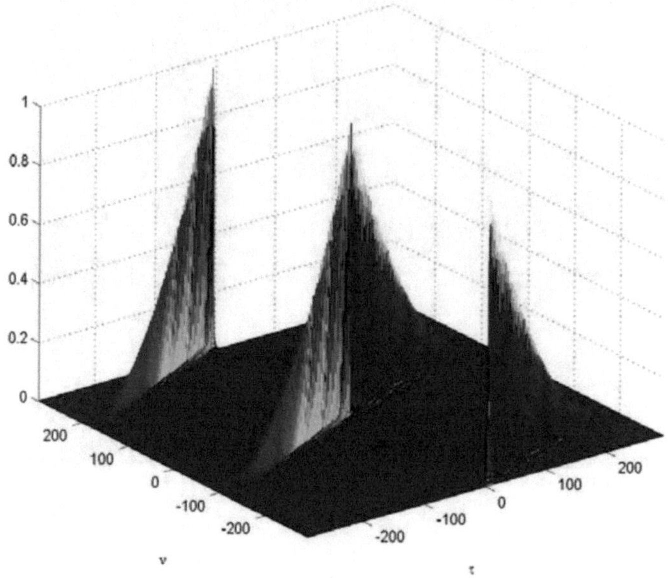

Figure 2.3: AF of a LFM sequence with $M = 293$.

Björck CAZAC waveforms and derive their AF surfaces for use in radar tracking to improve tracking performance.

A cyclically permuted Björck CAZAC sequence is given by $s(m)e^{\frac{j2\pi m\kappa}{M}}$ where κ is the frequency shift. We combine Q cyclically permuted Björck CAZACs,

$$s_q(m) = s(m)e^{\frac{j2\pi m\kappa_q}{M}} \tag{2.3}$$

where $q = 0, \ldots, Q-1$ using the multicarrier phase coded scheme. MCPC waveforms [1] are constructed by adding multiple waveforms that are modulated by orthogonal carriers separated in frequency using orthogonal frequency division multiplexing (OFDM). Each of the Q sequences is transmitted on a different carrier, with frequency separation between each carrier equal to $1/T_b$, where T_b is the sampling period. This increases the bandwidth of the waveform to Q/T_b (see Figure 2.5).

The MCPC CAZAC waveform, modulated with carrier frequency κ_c, is given by

$$g_\Theta(m) = \sum_{q=0}^{Q-1} s_q(\lfloor \frac{m}{Q} \rfloor) e^{-\frac{j2\pi qm}{Q}} e^{\frac{j2\pi m\kappa_c}{QM}} \tag{2.4}$$

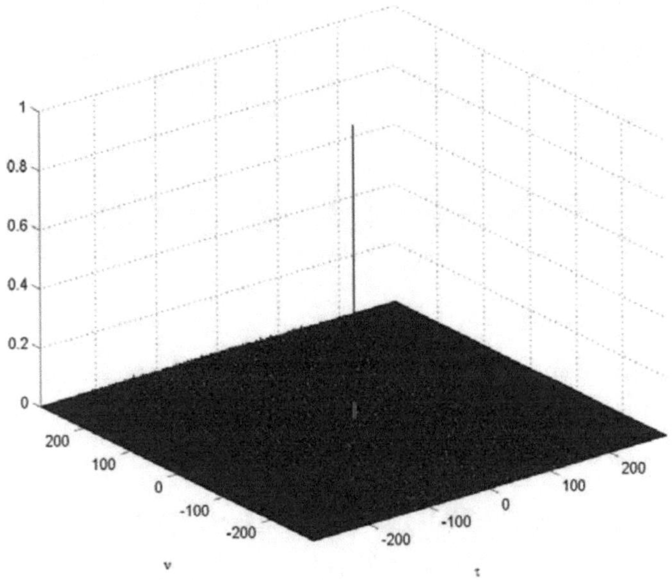

Figure 2.4: AF of a Björck CAZAC sequence with $M = 293$.

where the notation $\lfloor . \rfloor$ denotes rounding down to the nearest integer and $m = 0, \ldots, MQ - 1$. Note that the sampling rate of the MCPC CAZAC and is increased by Q as compared with the sampling rate of the single-carrier phase coded (SCPC) CAZAC. The vector $\Theta = (Q, M, \kappa)$ defines the parameters of the MCPC waveform. We choose to restrict $\kappa_q = \kappa q$ where $\kappa = 0, \ldots, M - 1$. This selection of cyclic frequency shifts causes the positioning of the sidelobes of the AF to depend on κ and, thus, facilitates the adaptive configuration of the transmitted waveform as we will demonstrate next.

The narrowband assumption used in (2.2) also holds for MCPC CAZACs. Specifically, we construct MCPC sequences with Q Björck CAZAC sequences whose length is approximately Q times smaller than the length of the single Björck CAZAC sequence. Therefore, we restrict the time-bandwidth product of the MCPC sequence to be the same as the time-bandwidth product of the single Björck CAZAC sequence. Note also that we double the number of possible AFs configurations by taking the Fourier transform (FT) of each of the MCPC waveforms that we construct. The AF of the transformed waveform is equal to the AF of the original signal with the delay and Doppler variables interchanged. This offers a convenient method of producing additional sidelobe positioning options with little effort.

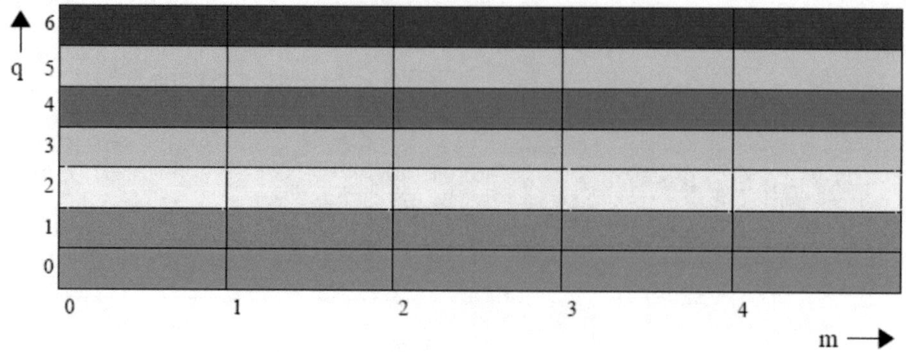

Figure 2.5: Construction of an MCPC waveform from $Q = 6$ CAZACs of length $M = 5$. Each row represents a different SCPC CAZAC with a different carrier frequency q and indices in time represented by m.

2.3.1 AF SURFACE OF MCPC BJÖRCK CAZAC SEQUENCES

The AF surface of the MCPC Björck CAZAC is given by $\mathcal{A}_{g_\Theta}(\tau, \nu) = |\mathrm{AF}_{g_\Theta}(\tau, \nu)|^2$

$$\mathrm{AF}_{g_\Theta}(\tau, \nu) = \frac{1}{E_{g_\Theta}} \sum_{m=0}^{MQ-1} g_\Theta(m-\tau) e^{\frac{j2\pi m \nu}{QM}} g_\Theta^*(m)$$

where $E_{g_\Theta} = \sum_{m=0}^{MQ-1} g_\Theta(m) g_\Theta^*(m)$ is the energy of $g_\Theta(m)$, which is normalized to have the same energy as $s(m)$. Using (2.3) and (2.4) the AF surface becomes

$$\mathrm{AF}_{g_\Theta}(\tau, \nu) = \frac{1}{E_{g_\Theta}} \sum_{m=0}^{MQ-1} s(\lfloor \frac{m-\tau}{Q} \rfloor) e^{\frac{j2\pi m \nu}{QM}} s^*(\lfloor \frac{m}{Q} \rfloor) \sum_{q=0}^{Q-1} e^{\frac{j2\pi \lfloor \frac{m-\tau}{Q} \rfloor (\kappa q)_M}{M}} e^{-j2\pi \frac{q(m-\tau)}{Q}}$$
$$\sum_{\hat{q}=0}^{Q-1} e^{-\frac{j2\pi \lfloor \frac{m}{Q} \rfloor (\kappa \hat{q})_M}{M}} e^{\frac{j2\pi \hat{q} m}{Q}} \tag{2.5}$$

We consider next two separate cases: $\kappa = 0$ and $\kappa > 0$.

Zero cyclic frequency-shift: With frequency shift $\kappa = 0$ and thus $\Theta = (Q, M, 0)$, then

$$\mathrm{AF}_{g_\Theta}(\tau, \nu) = \frac{1}{E_{g_\Theta}} \sum_{m=0}^{MQ-1} s(\lfloor \frac{m-\tau}{Q} \rfloor) e^{\frac{j2\pi m \nu}{QM}} s^*(\lfloor \frac{m}{Q} \rfloor) \sum_{q=0}^{Q-1} e^{-\frac{j2\pi q(m-\tau)}{Q}} \sum_{\hat{q}=0}^{Q-1} e^{\frac{j2\pi \hat{q} m}{Q}}.$$

Using the fact that the term $\sum_{\hat{q}=0}^{Q-1} e^{\frac{j2\pi \hat{q} m}{Q}} = Q\delta(m - \tilde{m}Q)$ with $\tilde{m}$ an integer, taking values as $\tilde{m} = 0, \ldots, M-1$, and replacing m with $\tilde{m}Q$ in the above expression, we have that

$$\mathrm{AF}_{g_\Theta}(\tau,\nu) = \frac{1}{E_{g_\Theta}} Q \sum_{\tilde{m}=0}^{M-1} s(\lfloor \frac{\tilde{m}Q-\tau}{Q} \rfloor) e^{\frac{j2\pi\tilde{m}\nu}{M}} s^*(\tilde{m}) \sum_{q=0}^{Q-1} e^{\frac{j2\pi q\tau}{Q}}.$$

Similarly, next we use the fact that $\sum_{q=0}^{Q-1} e^{\frac{j2\pi q\tau}{Q}} = Q\delta(\tau - \tilde{\tau}Q)$ where we define $\tilde{\tau}$ to be an integer. Furthermore, replacing τ with $\tilde{\tau}Q$, we obtain

$$\mathrm{AF}_{g_\Theta}(\tau,\nu) = \frac{1}{E_{g_\Theta}} Q^2 \sum_{\tilde{m}=0}^{M-1} s(\tilde{m}-\tilde{\tau}) e^{\frac{j2\pi\tilde{m}\nu}{M}} s^*(\tilde{m}). \tag{2.6}$$

From the above, it is clear that AF surface values in (2.6) are non-zero only for τ multiple of Q. Thus, zero AF surface regions exist of width Q. Although these regions can be used to reveal weak targets at selected areas in the AF surface plane, we also need to reduce the sidelobes near the origin of the AF. The area in the delay-Doppler measurement space near the AF origin is the area that is most commonly interrogated by the IPLPF tracker when accurately tracking a target as described in Chapter 5. Since we already have zero AF surface regions in the interval $\tau = 1, \ldots, Q-1$, we need to investigate the shape of the AF surface along the Doppler axis ν at $\tau = 0$. Setting $\tilde{\tau} = 0$ in (2.6), we have

$$\mathrm{AF}_{g_\Theta}(0,\nu) = \frac{1}{E_{g_\Theta}} Q^2 \sum_{\tilde{m}=0}^{M-1} s(\tilde{m}) e^{\frac{j2\pi\tilde{m}\nu}{M}} s^*(\tilde{m}).$$

Since $\tilde{m} = 0, \ldots, M-1$ and $|s(m)| = 1$ for all m, we conclude that the AF surface is non-zero only at ν integer multiple of M. Therefore, the location of sidelobes when $\tau = 0$ can also be chosen by adjusting the value of M. An example of this is shown in Figure 2.6 that depicts the AF surface of the MCPC waveform with $\Theta = (130, 13, 0)$. As it can be seen, all non-zero sidelobes exist at τ integer multiple of Q.

Positive cyclic frequency-shift: Next, we consider the case of $\kappa > 0$, that, as we will see, allows us to have more diversity in the locations of the AF sidelobes. For convenience, we repeat (2.5) below.

$$\mathrm{AF}_{g_\Theta}(\tau,\nu) = \frac{1}{E_{g_\Theta}} \sum_{m=0}^{MQ-1} s(\lfloor \frac{m-\tau}{Q} \rfloor) e^{\frac{j2\pi m\nu}{QM}} s^*(\lfloor \frac{m}{Q} \rfloor)$$

$$\sum_{q=0}^{Q-1} e^{\frac{j2\pi \lfloor \frac{m-\tau}{Q} \rfloor (\kappa q)_M}{M}} e^{-\frac{j2\pi q(m-\tau)}{Q}} \sum_{\hat{q}=0}^{Q-1} e^{-\frac{j2\pi \lfloor \frac{m}{Q} \rfloor (\kappa\hat{q})_M}{M}} e^{\frac{j2\pi\hat{q}m}{Q}}.$$

For $\kappa \neq 0$, we observe that terms $e^{\frac{j2\pi \lfloor \frac{m-\tau}{Q} \rfloor (\kappa q)_M}{M}}$ and $e^{-\frac{j2\pi \lfloor \frac{m}{Q} \rfloor (\kappa\hat{q})_M}{M}}$ repeat a number of times in the above summations. Therefore, we can factor these terms and rewrite (2.5) as:

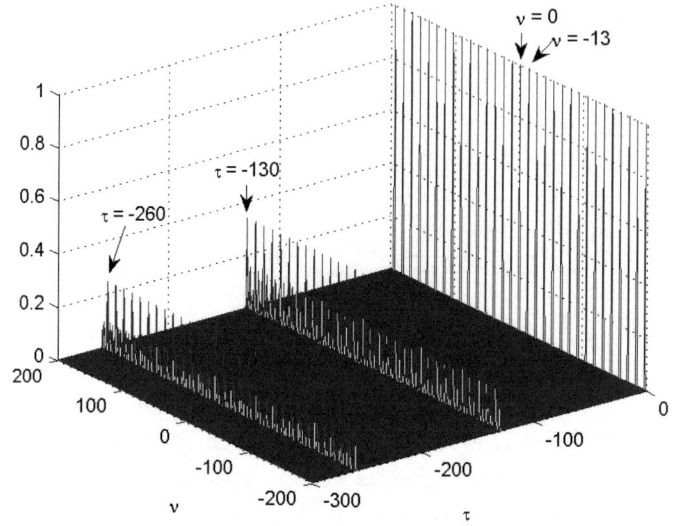

Figure 2.6: AF surface plot of an MCPC Björck CAZAC with parameters $\Theta = (Q, M, \kappa) = (130, 13, 0)$.

$$\mathrm{AF}_{g_\Theta}(\tau, \nu) = \frac{1}{E_{g_\Theta}} \sum_{m=0}^{MQ-1} s(\lfloor \frac{m-\tau}{Q} \rfloor) e^{\frac{j2\pi m\nu}{QM}} s^*(\lfloor \frac{m}{Q} \rfloor)$$

$$\sum_{\alpha=0}^{\beta-1} e^{\frac{j2\pi \lfloor \frac{m-\tau}{Q} \rfloor \kappa\alpha}{M}} \sum_{q=0}^{\lfloor \frac{Q}{\beta} \rfloor - 1} e^{-\frac{j2\pi(\beta q+\alpha)(m-\tau)}{Q}} \sum_{\hat{\alpha}=0}^{\beta-1} e^{-\frac{j2\pi \lfloor \frac{m}{Q} \rfloor \kappa\hat{\alpha}}{M}} \sum_{\hat{q}=0}^{\lfloor \frac{Q}{\beta} \rfloor - 1} e^{\frac{j2\pi(\beta\hat{q}+\hat{\alpha})m}{Q}} .$$

Note that q and $\hat{q}$ now vary from 0 to $\lfloor \frac{Q}{\beta} \rfloor - 1$. We choose Q, M and κ such that $\beta = \lfloor \frac{M-1}{\kappa} \rfloor + 1$ is approximately a multiple of Q for most of choices of $\kappa = 1, \dots, M-1$. This eliminates terms with q such that $\beta(\lfloor \frac{Q}{\beta} \rfloor - 1) + \beta - 1 \leq q \leq Q - 1$, and similarly for $\hat{q}$, that were omitted in the above expression. These terms would cause a small distortion in the AF surface that we would like to ideally have and is described below. The expression for the AF surface further becomes the following:

$$\mathrm{AF}_{g_\Theta}(\tau, \nu) = \frac{1}{E_{g_\Theta}} \sum_{m=0}^{MQ-1} s(\lfloor \frac{m-\tau}{Q} \rfloor) e^{\frac{j2\pi m\nu}{QM}} s^*(\lfloor \frac{m}{Q} \rfloor)$$

$$\sum_{\alpha=0}^{\beta-1} e^{\frac{j2\pi\lfloor\frac{m-\tau}{Q}\rfloor\kappa\alpha}{M}} e^{-\frac{j2\pi\alpha(m-\tau)}{Q}} \sum_{q=0}^{\lfloor\frac{Q}{\beta}\rfloor-1} e^{-\frac{j2\pi q(m-\tau)}{Q/\beta}} \sum_{\hat{\alpha}=0}^{\beta-1} e^{-\frac{j2\pi\lfloor\frac{m}{Q}\rfloor\kappa\hat{\alpha}}{M}} e^{\frac{j2\pi\hat{\alpha}m}{Q}} \sum_{\hat{q}=0}^{\lfloor\frac{Q}{\beta}\rfloor-1} e^{\frac{j2\pi\hat{q}m}{Q/\beta}}.$$

Using the fact that the term $\sum_{\hat{q}=0}^{\lfloor\frac{Q}{\beta}\rfloor-1} e^{\frac{j2\pi\hat{q}m}{Q/\beta}} = \frac{Q}{\beta}\delta(m-\tilde{m}\frac{Q}{\beta})$ with $\tilde{m}$ an integer, taking values as $\tilde{m}=0,\ldots,\beta M-\frac{\beta}{Q}$, and if we let $\tilde{m}=\hat{m}Q+\check{m}\frac{Q}{\beta}$, $0\le\hat{m}\le M-1$, $0\le\check{m}\le\beta-1$ in the above expression, we have that

$$\mathrm{AF}_{g_\Theta}(\tau,\nu) = \frac{1}{E_{g_\Theta}}\frac{Q}{\beta}\sum_{m=0}^{MQ-1} s(\lfloor\frac{\hat{m}Q+\check{m}\frac{Q}{\beta}-\tau}{Q}\rfloor)e^{\frac{j2\pi(\beta\hat{m}+\check{m})\nu}{\beta M}} s^*(\hat{m})$$

$$\sum_{\alpha=0}^{\beta-1} e^{\frac{j2\pi\lfloor\frac{\hat{m}Q+\check{m}\frac{Q}{\beta}-\tau}{Q}\rfloor\kappa\alpha}{M}} e^{-\frac{j2\pi\alpha(\hat{m}Q+\check{m}\frac{Q}{\beta}-\tau)}{Q}} \sum_{q=0}^{\lfloor\frac{Q}{\beta}\rfloor-1} e^{\frac{j2\pi q\tau}{Q/\beta}} \sum_{\hat{\alpha}=0}^{\beta-1} e^{-\frac{j2\pi\hat{m}\kappa\hat{\alpha}}{M}} e^{\frac{j2\pi\hat{\alpha}\check{m}}{\beta}}.$$

Similarly, we use the fact that $\sum_{q=0}^{\lfloor\frac{Q}{\beta}\rfloor-1} e^{\frac{j2\pi q\tau}{Q/\beta}} = \frac{Q}{\beta}\delta(\tau-\tilde{\tau}\frac{Q}{\beta})$ where we define $\tilde{\tau}$ to be an integer. Furthermore, we let $\tau=\hat{\tau}Q+\check{\tau}\frac{Q}{\beta}$ with $\hat{\tau}$ and $\check{\tau}$ integers and $0\le\check{\tau}\le\beta-1$, we have that

$$\mathrm{AF}_{g_\Theta}(\tau,\nu) = \frac{1}{E_{g_\Theta}}\frac{Q^2}{\beta^2}\sum_{\hat{m}=0}^{M-1}\sum_{\check{m}=0}^{\beta-1} s(\hat{m}-\hat{\tau}+\lfloor\frac{\check{m}-\check{\tau}}{\beta}\rfloor)e^{\frac{j2\pi(\hat{m}\beta+\check{m})\nu}{\beta M}} s^*(\hat{m})$$

$$\sum_{\alpha=0}^{\beta-1} e^{\frac{j2\pi(\hat{m}-\hat{\tau}+\lfloor\frac{\check{m}-\check{\tau}}{\beta}\rfloor)\kappa\alpha}{M}} e^{-\frac{j2\pi(\check{m}-\check{\tau})\alpha}{\beta}} \sum_{\hat{\alpha}=0}^{\beta-1} e^{-\frac{j2\pi\hat{m}\kappa\hat{\alpha}}{M}} e^{\frac{j2\pi\check{m}\hat{\alpha}}{\beta}} \tag{2.7}$$

The above expression shows that non-zero values of the AF surface exist for $\hat{\tau}$ integer and $0\le\check{\tau}\le\beta-1$, i.e., $\tau=\tilde{\tau}\frac{Q}{\beta}$, where $\tilde{\tau}$ is an integer. This provides valleys in the AF surface space with controlled sizes.

We also examine what happens along the Doppler axis ν at zero delay and for $\kappa>0$. We set $\tau=0$ or $\hat{\tau}=0$ and $\check{\tau}=0$ in (2.7). We have that

$$\mathrm{AF}_{g_\Theta}(0,\nu) = \frac{1}{E_{g_\Theta}}\frac{Q^2}{\beta^2}\sum_{\hat{m}=0}^{M-1}\sum_{\check{m}=0}^{\beta-1} s(\hat{m})s^*(\hat{m})e^{\frac{j2\pi(\hat{m}\beta+\check{m})\nu}{\beta M}}$$

$$\sum_{\alpha=0}^{\beta-1} e^{\frac{j2\pi\hat{m}\kappa\alpha}{M}} e^{-\frac{j2\pi\check{m}\alpha}{\beta}} \sum_{\hat{\alpha}=0}^{\beta-1} e^{-\frac{j2\pi\hat{m}\kappa\hat{\alpha}}{M}} e^{\frac{j2\pi\check{m}\hat{\alpha}}{\beta}}.$$

The terms $e^{-\frac{j2\pi(\hat{m}\beta+\check{m})\nu}{\beta M}}$ reveal that the AF has sidelobes periodically repeating with period βM on the Doppler axis at $\tau=0$. Evaluating the above expression at the in-between intervals, we can easily obtain the AF surface sidelobe values. We then choose to use only waveforms with parameters Q, M, and κ with relatively low sidelobe levels in their AF surface with $\tau = 0$.

In another special case when $\kappa = 1$, which implies that $\beta = M$, we have larger valleys. The AF surface in (2.7) becomes the following:

$$\mathrm{AF}_{g_\Theta}(\tau, \nu) = \frac{1}{E_{g_\Theta}} \frac{Q^2}{\beta^2} \sum_{\hat{m}=0}^{M-1} \sum_{\check{m}=0}^{\beta-1} s(\hat{m} - \hat{\tau} + \lfloor \frac{\check{m} - \check{\tau}}{\beta} \rfloor) e^{j2\pi \frac{(\hat{m}\beta + \check{m})\nu}{\beta^2}} s^*(\hat{m})$$

$$\sum_{\alpha=0}^{\beta-1} e^{\frac{j2\pi(\hat{m} - \check{m} - (\hat{\tau} - \check{\tau}) + \lfloor \frac{\check{m} - \check{\tau}}{\beta} \rfloor)\alpha}{\beta}} \sum_{\tilde{\alpha}=0}^{\beta-1} e^{-j2\pi \frac{(\hat{m} - \check{m})\tilde{\alpha}}{\beta}}.$$

We use the fact that $\sum_{\tilde{\alpha}=0}^{\beta-1} e^{-j2\pi(\hat{m}-\check{m})\tilde{\alpha}/\beta} = \beta\delta(\hat{m} - \check{m})$ since $0 \leq \hat{m} \leq M - 1, 0 \leq \check{m} \leq \beta - 1$. Therefore, we let $m = \hat{m} = \check{m}$ in the AF expression above. Then we note that the factor $\lfloor (m - \check{\tau})/\beta \rfloor$ can only take the values of 0 if $m = \check{\tau}$ and -1 if $m < \check{\tau}$ since both m and $\check{\tau}$ take values less than β. This implies that non-zero values of the AF surface $\mathcal{A}_{g_\Theta}(\tau, \nu)$ only exist at delay locations such that $\hat{\tau} - \check{\tau}$ or $\hat{\tau} - \check{\tau} + 1$ are multiples of β. For $\lfloor (m - \check{\tau})/\beta \rfloor = -1$, which restricts $\hat{\tau} - \check{\tau} + 1$ to be a multiple of β with $m < \check{\tau}$, we observe that there is little overlap of non-zero values between $s(m - \hat{\tau} + \lfloor (m - \check{\tau})/\beta \rfloor)$ and $s^*(m)$. Therefore, we can neglect the case where $\lfloor (m - \check{\tau})/\beta \rfloor = -1$ in the AF expression. Letting $\lfloor (m - \check{\tau})/\beta \rfloor = 0$, and using $\sum_{\alpha=0}^{\beta-1} e^{-j2\pi(\hat{\tau} - \check{\tau})\alpha/\beta} = \beta\, \delta(\hat{\tau} - \check{\tau})$, we can obtain

$$\mathrm{AF}_{g_\Theta}(\tau, \nu) \approx \frac{1}{E_{g_\Theta}} Q^2 \sum_{\hat{m}=0}^{M-1} s(\hat{m} - \hat{\tau}) e^{\frac{j2\pi(\hat{m}\beta + \hat{m})\nu}{\beta^2}} s^*(\hat{m}) \delta(\hat{\tau} - \check{\tau}).$$

Therefore, $\hat{\tau} = \check{\tau}$. Since $\tau = \hat{\tau} Q + \check{\tau} \frac{Q}{\beta}$ with $\beta = M$ then non-zero sidelobes of the AF surface appear at intervals of $Q + \frac{Q}{M}$ in the delay. The AF surface of the MCPC waveform with $\Theta = (130, 13, 1)$ is shown in Figure 2.7.

In summary, the possibility to choose the parameters $\Theta = (Q, M, \kappa)$ of an MCPC Björck CAZAC waveform and also to rotate the entire AF surface by choosing to take the FT of the waveform enables us to position sidelobes in order to minimize the predicted mean squared error as shown in Chapter 5.

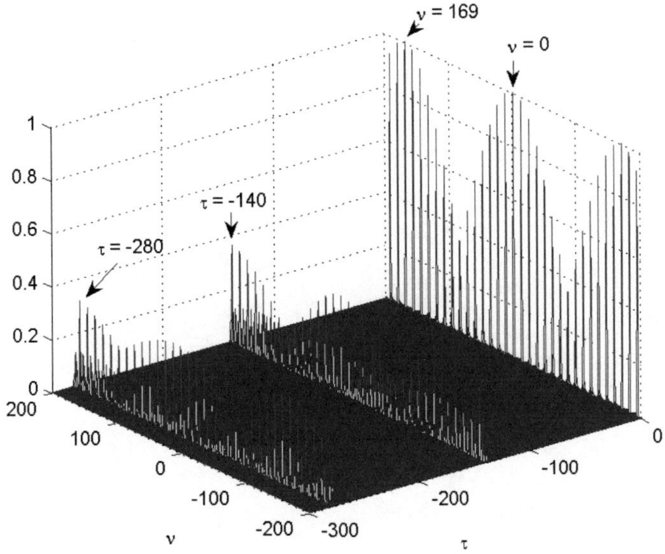

Figure 2.7: AF surface plot of an MCPC Björck CAZAC with parameters $\Theta = (Q, M, \kappa) = (130, 13, 1)$.

CHAPTER 3

Target Tracking with a Particle Filter

In this chapter, we provide an overview of commonly used tracking methods together with an introduction to Bayesian inference and the particle filtering concept. We also describe particle filtering methods for tracking a single and multiple targets. Finally, we provide the basics of radar tracking.

3.1 TRACKING METHODS AND MODELS

The estimation of target state elements such as position and velocity finds use in many applications [23, 20, 21, 22, 19]. Tracking can be performed in three ways: smoothing, estimation and prediction. In smoothing, an available set of data is used to obtain an estimate of values at a time prior to the latest measurement. In estimation, an estimate is obtained utilizing measurements up to the current time of the estimate. Finally, prediction is used to obtain an estimate at future time steps, based on previous data and any other kind of information available.

The most common approach in target tracking problems is a probabilistic one. In this approach, the kinematics of the target under investigation are given by a motion model that describes the evolution of the state of a target. Using the Bayesian method, a belief on the state of the target given some observation is found using the posterior probability density function (pdf) that is calculated recursively at each time step. An optimal solution to this tracking problem is offered through the Kalman filter [24, 25] that follows the steps of prediction and update. This solution is attainable only when the motion and observation models are assumed known and linear, and when the process and observation noise sequence are assumed Gaussian. The assumptions of linearity and Gaussianity, however, are very restrictive. Thus, in many situations, the Kalman filter fails to provide adequate tracking. Some approximations can be used to modify Kalman filtering for these difficult scenarios [4]. Other suboptimal solutions are the grid-based methods that offer recursion of the posterior, assuming that the state space is discrete. Grid-based methods can become computationally intractable as the state space dimension increases and fail in difficult scenarios and multiple target tracking situations.

Some suboptimal methods exist, however, that relax the strict assumptions of the above methods. The extended Kalman filter (EKF) [26] uses local linearization techniques to describe non-linear motion and observation models. Since the posterior is approximated by a Gaussian, the Kalman recursion can still be implemented. The unscented Kalman filter (UKF) [27], approximates nonlinearities better by using the unscented transform, which in this case calculates the statistics of Gaussian random variables propagated through non-linear systems.

Approximate grid-based methods [4] can approximate non-Gaussian posterior densities when the continuous state space can be decomposed into cells. As the state space increases, however, there exists a prohibitive computational cost. Finally, particle filtering methods [4] require no linearity or Gaussianity assumptions and the computational complexity associated with them is tractable and thus, are becoming increasingly popular.

The particle filter requires a kinematic and a measurement model which are explained below. Kinematic models deal with the evolution of the state of one or many targets. These models are mainly based on the nature of the moving objects. A state vector $\mathbf{x}_k$, usually comprising of position and velocity, describes the state. The state vector evolves as

$$\mathbf{x}_k = \mathbf{f}_k(\mathbf{x}_{k-1}, \mathbf{v}_{k-1}) \tag{3.1}$$

where $\mathbf{f}_k : \Re^{e_x} \times \Re^{e_v} \rightarrow \Re^{e_x}$, with e_x and e_v denoting the dimension of $\mathbf{x}$ and $\mathbf{v}$ respectively, is a function of $\mathbf{x}_{k-1}$ and can either be linear or non-linear. $\mathbf{v}_{k-1}$ is an independent and identically distributed (i.i.d) process noise sequence. are the dimensions of the state and process vectors respectively.

Another component of Bayesian estimation is the measurement model that characterizes the relationship between the observed data and the state of the targets. The measurement model is represented as

$$\mathbf{y}_k = \mathbf{h}_x(\mathbf{x}_k, \mathbf{n}_k) \tag{3.2}$$

where $\mathbf{h}_k : \Re^{e_x} \times \Re^{e_n} \rightarrow \Re^{e_y}$ is a known linear or non-linear function of the state $\mathbf{x}_k$ and the i.i.d measurement sequence $\mathbf{n}_k$.

Examples of measurements include range measurements that provide the distance of the target to the sensor, range rate measurements that measure the rate of change of the distance between the target and sensor, azimuth and elevation angles between the target and sensor in the horizontal and vertical direction, respectively, and sequential images of the position of the target. Some types of sensors often used for target tracking are radar, infrared, acoustic and optical.

3.2 BAYESIAN INFERENCE

In order to provide the framework in which the particle filter operates, it is necessary to provide the basics of Bayesian inference. Bayesian inference consists of finding the likelihood function and combining it with the prior information using Bayes theorem to compute a posterior distribution [28]. This will provide a Bayesian recursion useful in many applications.

Let $\mathbf{x}_{1:K} = [\mathbf{x}_1, \ldots, \mathbf{x}_K]$, where the random vectors $\mathbf{x}_k$ with time step $k = 1, \ldots, K$, take values in a state space $\Re^n$. Information about $\mathbf{x}_{1:K}$ is obtained through observations $\mathbf{y}_{1:K} = [\mathbf{y}_1, \ldots, \mathbf{y}_K]$, with each $\mathbf{y}_k$ being within a measurement space $\Re^m$. The likelihood function $p(\mathbf{y}_k|\mathbf{x}_k)$ is obtained using the measurement model as in (3.2). With the Bayesian perspective, it is possible, in two stages, to recursively obtain the posterior pdf, $p(\mathbf{x}_k|\mathbf{y}_{1:K})$, which represents a belief in the state $\mathbf{x}_k$ at time k based on the measurements up to time k. The two stages are prediction and update.

In the prediction step, it is assumed that the pdf $p(\mathbf{x}_{k-1}|\mathbf{y}_{1:k-1})$ at time $k-1$ is available. Using the kinematic prior, $p(\mathbf{x}_k|\mathbf{x}_{k-1})$, coming from the motion model given by (3.1), the posterior $p(\mathbf{x}_k|\mathbf{y}_{1:k-1})$ is obtained, via the Chapman-Kolmogorov equation

$$p(\mathbf{x}_k|\mathbf{y}_{1:k-1}) = \int p(\mathbf{x}_k|\mathbf{x}_{k-1})p(\mathbf{x}_{k-1}|\mathbf{y}_{1:k-1})d\mathbf{x}_{k-1}. \tag{3.3}$$

When a measurement $\mathbf{y}_k$ becomes available at time k, the update stage follows, where the prior is updated using Bayes theorem:

$$p(\mathbf{x}_k|\mathbf{y}_{1:k}) = \frac{p(\mathbf{y}_k|\mathbf{x}_k)p(\mathbf{x}_k|\mathbf{y}_{1:k-1})}{\int p(\mathbf{y}_k|\mathbf{x}_k)p(\mathbf{x}_k|\mathbf{y}_{1:k-1})d\mathbf{x}_k}. \tag{3.4}$$

The recursive method described above constitutes the Bayesian inference and shows the ability of this method to accommodate new data in updating information about the state $\mathbf{x}_k$. Different likelihood functions exist for different kinds of sensors, and the Bayesian inference can be built for many different kinds of motion models.

3.3 THE PARTICLE FILTER ALGORITHM

Monte Carlo methods are a family of methods that have the common characteristic of solving problems by generating a set of random numbers, based on an educated guess, and calculating the correctness of the guess based on standards given by the problem. Therefore, the Monte Carlo approach is used for problems for which it is not possible to find analytical solutions.

A specific example that proves very useful in tracking problems is the approximation of a pdf by discrete samples. When possible, independent and identically distributed samples are drawn from the target density. Wherever the pdf is peaked, more samples are gathered at that point due to higher probability of occurrence. Therefore, the value of the pdf at discrete points can be estimated by the quantity of samples at that point. The greater the number of samples, the better the density is characterized.

The procedure is, of course, computationally expensive as it requires an adequate amount of discrete samples for representing the density at hand. Through the advancements in computer technology, however, this problem was alleviated.

Sequential Monte Carlo (SMC) methods are used to solve estimation problems by sequentially updating knowledge about the estimates, based on sequentially arriving data. An example of an SMC method is the sequential importance sampling (SIS) [29, 30]; this is also known as particle filtering [31], bootstrap filtering [32], the condensation algorithm [33], interacting particle approximations [34, 35], and survival of the fittest [36].

The main reason for using the SMC technique is to solve problems defined by non-linear, non-Gaussian scenarios, thus, removing the limitiations imposed by Kalman filtering. The hard to compute posterior density can be approximated by a large set of point estimates that are given weights according to their correctness in describing the density.

The particle filter approximates the posterior density $p(\mathbf{x}_k|\mathbf{y}_k)$ as:

$$p(\mathbf{x}_k|\mathbf{y}_k) \approx \sum_{n=1}^{N} w_k^n \delta(\mathbf{x}_k - \mathbf{x}_k^n) \tag{3.5}$$

where w_k^n is the weight assigned to particle $\mathbf{x}_k^n$, n is the particle index, $n = 1, \ldots, N$, and N is the number of particles used. The particles $\mathbf{x}_k^n$, are from an appropriate importance density $q(\mathbf{x}_k|\mathbf{x}_{k-1}, \mathbf{y}_k)$ [29, 37]. The optimal importance density has been found to be the one that incorporates information from the kinematics of previous time steps as well as current observations. However, the use of an exact or approximately optimal importance density can be a hard task to perform. If it is not possible to use an importance density that incorporates the current measurements $\mathbf{y}_k$, then the kinematic prior $p(\mathbf{x}_k|\mathbf{x}_{k-1})$ is used as the importance density. The weights of the particles are updated using the weight equation which utilizes the measurement likelihood, the kinematic prior and the importance density

$$w_k^n \propto w_{k-1}^n \frac{p(\mathbf{y}_k|\mathbf{x}_k^n)\, p(\mathbf{x}_k^n|\mathbf{x}_{k-1}^n)}{q(\mathbf{x}_k^n|\mathbf{x}_{k-1}^n, \mathbf{y}_k)} \tag{3.6}$$

which reduces to

$$w_k^n \propto w_{k-1}^n p(\mathbf{y}_k|\mathbf{x}_k^n) \tag{3.7}$$

if the kinematic prior is used as the importance density.

Using the particle filter posterior density approximation in (3.5) an estimate of the expected value of any function of the state $g(\mathbf{x}_k)$ can be obtained as

$$\mathbf{E}[g(\mathbf{x}_k)] = \int_{\mathbf{x}_k} g(\mathbf{x}_k) p(\mathbf{x}_k|\mathbf{y}_k) d\mathbf{x}_k \approx \sum_{n=1}^{N} w_k^n g(\mathbf{x}_k^n). \tag{3.8}$$

By increasing the number of particles, the SIS algorithm approaches optimality. By doing so, however, the computational expense of the method increases, especially when dealing with multiple targets. A good choice of importance density usually enables a reduction in the number of particles required.

3.3.1 THE DEGENERACY EFFECT

A problem associated with particle filtering is that the variance of the importance weights increases over time [29]. This implies that only few particles will receive significant weights and thus will not represent the posterior density adequately. Therefore, a large computational effort will be allocated in propagating incorrect samples, with the additional drawback of the few good particles being insufficient to represent the posterior. This is referred to as the particle degeneracy problem [4].

Increasing the number of particles reduces the effect of degeneracy to some extent but has the obvious computational drawbacks. Choosing a good importance density and the use of resampling [4, 38] can reduce the degeneracy effect, as discussed next.

3.3.2 CHOICE OF IMPORTANCE DENSITY

The optimal importance density is one that enables us to propose new particles based on both the previous state and the current measurements and is described by $p(\mathbf{x}_k|\mathbf{x}_{k-1}, \mathbf{y}_k)$. This, however, is possible only in particular circumstances: when the state space is a member of a finite set, or when the dynamics are Gaussian [4]. In most cases encountered, however, this is not true; therefore, other methods or densities need to be used. A local approximation of the optimal importance density can be obtained with the use of local linearization techniques [29]. A very simple to use importance density is the kinematic prior, $p(\mathbf{x}_k|\mathbf{x}_{k-1})$, which is easy to sample from and its use reduces the weight equation on being dependent on the measurement likelihood. This, however, often causes many particles to be generated that do not represent the posterior adequately, especially when the measurement likelihood is significantly more peaked than the kinematic prior. By the use of the optimal importance function or its approximation, the selection of the particles is biased towards the data, therefore achieving lower particle rejection rates and, consequently, the reduction in the number of particles needed to perform successful tracking.

3.3.3 RESAMPLING

Degeneracy can be solved by the use of resampling, by which the SIS becomes the sequential importance resampling (SIR) particle filter. In this method, the particles that have small weights are eliminated, and the particles with large weights, and that represent the posterior better are propagated to the next step. The particles that survive the resampling procedure are then given equal weights. This procedure can be applied whenever N_{eff} falls below a certain threshold. Different resampling schemes exist. The most preferred method is systematic resampling [39], while other popular schemes are stratified and residual sampling methods [40]. The sampling importance resampling (SIR) single particle filter algorithm using systematic resampling is given in Table 3.1.

Table 3.1: SIR Single Target Tracking Algorithm

- For each particle $n = 1, \dots, N$
  - Draw particle $\mathbf{x}_k^n \sim q(\mathbf{x}_k|\mathbf{x}_{k-1}^n, \mathbf{y}_k)$
- Compute the weight of each particle as $w_k^n \propto w_{k-1}^n \frac{p(\mathbf{y}_k|\mathbf{x}_k^n)p(\mathbf{x}_k^n|\mathbf{x}_{k-1}^n)}{q(\mathbf{x}_k^n|\mathbf{x}_{k-1}^n,\mathbf{y}_k)}$
- Sample N particles with replacement from the distribution of w_k

3.4 TRACKING MULTIPLE TARGETS: THE INDEPENDENT PARTITION PARTICLE FILTER

In order to track multiple targets, the SIR single target particle filter [4] extends directly to the SIR multitarget particle filter [41]. This filter proposes particles that consist of partitions, each representing one of the targets of interest. The partitions in a particle, however, are proposed simultaneously and placed into particles without any additional assessment of their validity. Therefore, many of the particles that are built receive low weights at the particle weighing step. For example, some of the partitions of the particle might be very good estimates of the state vector of certain targets, while other partitions in the same particle can be bad estimates of other target's state vectors. Thus, many particles can be composed of both good and bad estimates. This results in particles receiving an overall low weight, due to the bad partitions.

The IP algorithm [42], offers a method of constructing better particles. This is achieved by independently proposing partitions for each particle, weighing them and choosing the partitions that receive the best weights by sampling from the weight density of each partition. This constitutes the proposal subroutine. More specifically, the partitions are proposed via a single partition importance density of our choice and weighted by a function that contains useful information coming from the measurements. The chosen partitions are then placed into particles, and the resulting particles are weighted. As a further advantage, the overall better particles have greater chances of being propagated to the next time step, if resampling is used. This results in a great reduction in the number of particles required for representing the posterior adequately and an improvement in overall performance when compared to the SIR multitarget particle filter.

Usually an approximation of a density that involves single target measurement information is used as a weighing function in the proposal subroutine. A simple choice is an approximation to the single target measurement likelihood function. Note that each partition can be weighed individually because all the measurements are well separated in the sensor space. Then each partition will receive weights which are due only to the part of the joint posterior that represents the target corresponding to that partition. This is true even if, in practice, it is weighted against the whole posterior density. This is important as partitions can be proposed performing no data association.

Next, the assumptions that are used to implement the IP algorithm are presented more explicitly. The first assumption is that the dynamics of the particles are independent. Thus, the state transition density of the combination of the partitions, each representing the state of a targets, can be factorized as the following:

$$p(\mathbf{X}_k^n|\mathbf{X}_{k-1}^n) = \prod_{l=1}^{L} p(\mathbf{x}_{l,k}^n|\mathbf{x}_{l,k-1}^n). \tag{3.9}$$

If each of these factors are used as the single proposal density, it is possible to propose each partition of each particle, corresponding to a certain target, using the prior density of that target. Also, the joint posterior distribution can be factorized as

$$p(\mathbf{X}_k|\mathbf{Y}_k) = \prod_{l=1}^{L} p(\mathbf{x}_{l,k}|\mathbf{Y}_k) \,.$$

The above proceeds from the assumption that the targets are not coupled in the sensor space,

Table 3.2: Schematic of three particles, $n = 1, \ldots, 3$, with three partitions, $l = 1, \ldots 3$: $\mathbf{x}_{1,k}^n$, $\mathbf{x}_{2,k}^n$ and $\mathbf{x}_{3,k}^n$, where the time subscript k was omitted above. The particles are formed after independent proposal of individual partitions and sampling from the weight distribution of partitions. Here, the partitions shown on the left received good weights and were more likely to be chosen, thus forming particles that overall represent the posterior well.

Proposed Particles		Proposed particles after partition sampling
Particle 1 : $\{\mathbf{x}_1^1, \mathbf{x}_2^1, \mathbf{x}_3^1\}$		Particle 1 : $\{\mathbf{x}_1^3, \mathbf{x}_2^1, \mathbf{x}_3^2\}$
Particle 2 : $\{\mathbf{x}_1^2, \mathbf{x}_2^2, \mathbf{x}_3^2\}$	$\longrightarrow$	Particle 2 : $\{\mathbf{x}_1^1, \mathbf{x}_2^3, \mathbf{x}_3^2\}$
Particle 3 : $\{\mathbf{x}_1^3, \mathbf{x}_2^3, \mathbf{x}_3^3\}$		Particle 3 : $\{\mathbf{x}_1^1, \mathbf{x}_2^2, \mathbf{x}_3^3\}$

Since the joint posterior distribution of the multitarget state can be factorized into independent components, each corresponding to one of the partitions, then, for each particle, the partitions can be constructed independently. Therefore, partitions are proposed independently through the kinematic prior, $p(\mathbf{x}_{l,k}^n|\mathbf{x}_{l,k-1}^n)$, $l = 1, \ldots, L$, and eventually weighted. Therefore, the weight equation for each partition $l = 1, \ldots, L$ reduces to

$$w_{l,k}^n \propto p(\mathbf{Y}_k|\mathbf{x}_{l,k}^n) \,.$$

After resampling, that involves obtaining a sample j of the partition l from the distribution of $w_{l,k}$, the weight $w^j_{l,k}$ received by each partition $l = 1, \ldots, L$ becomes the bias $b^n_{l,k}$ for that partition, corresponding to particle n. Therefore, $b^n_{l,k}=w^j_{l,k}$.

With the IP, one may permute the order of partitions corresponding to the same targets between particles, constructing particles with good partitions. This is demonstrated in Table 3.2.

Another benefit of the IP technique is that the choice of the particles is biased towards the measurements. Since this is a biased sampling scheme, the bias of the partitions are retained and used in the final weighing step. To express the procedure in a mathematical form, a derivation of the weighing equation for the particles is presented. The weight of the particle $n = 1, \ldots, N$ will be:

$$w^n_k \propto w^n_{k-1} \frac{p(\mathbf{Y}_k|\mathbf{X}^n_k)p(\mathbf{X}^n_k|\mathbf{X}^n_{k-1})}{q(\mathbf{X}^n_k|\mathbf{X}^n_{k-1}, \mathbf{Y}_k)}$$

where $q(\mathbf{X}^n_k|\mathbf{X}^n_{k-1}, \mathbf{Y}_k)$ is the proposal density used to propose each of the particles. Recall that each of the partitions of the particle was proposed by the kinematic prior of the target it represents and weighted with the whole posterior. This weight biased its selection for placement in the particle. Therefore,

$$q(\mathbf{X}^n_k|\mathbf{X}^n_{k-1}, \mathbf{Y}_k) = \prod_{l=1}^{L} q(\mathbf{x}^n_{l,k}|\mathbf{x}^n_{l,k-1}, \mathbf{Y}_k) \tag{3.10}$$

and

$$q(\mathbf{x}^n_{l,k}|\mathbf{x}^n_{l,k-1}, \mathbf{Y}_k) = b^n_{l,k} p(\mathbf{x}^n_{l,k}|\mathbf{x}^n_{l,k-1}). \tag{3.11}$$

Combining (3.10) and (3.11), the weight of each particle becomes

$$w^n_k \propto w^n_{k-1} \frac{p(\mathbf{Y}_k|\mathbf{X}^n_k)p(\mathbf{X}^n_k|\mathbf{X}^n_{k-1})}{\prod_{l=1}^{L} b^n_{l,k} \prod_{l=1}^{L} p(\mathbf{x}^n_{l,k}|\mathbf{x}^n_{l,k-1})}$$

which, using (3.9), becomes

$$w^n_k \propto w^n_{k-1} \frac{p(\mathbf{Y}_k|\mathbf{X}^n_k)}{\prod_{l=1}^{L} b^n_{l,k}}.$$

The steps of the IP method are summarized in Table 3.3.

3.5 TRACKING WITH RADAR MEASUREMENTS

A typical target tracking radar system transmits a waveform and processes its return in order to extract information on the range and range rate of the target with respect to the radar sensors. The measurement accuracy in the range and range rate depends on the waveform used and the processing method of the return signal.

Table 3.3: Independent Partitions Algorithm [42]

- For each partition $l = 1, \ldots, L$
  - For each particle $n = 1, \ldots, N$
    - Sample $\mathbf{x}^n_{l,k} \sim p(\mathbf{x}_{l,k}|\mathbf{x}^n_{l,k-1})$
    - Compute $w^n_{l,k} = w^n_{l,k-1} p(\mathbf{Y}_k|\mathbf{x}^n_{l,k})$
  - Normalize $w_{l,k}$
  - Resample by choosing a partition j from the distribution of $w_{l,k}$ with replacement
  - Keep the bias of each sample as $b^n_{l,k} = w^j_{l,k}$
- For each particle $n = 1, \ldots, N$
  - Compute $w^n_k = w^n_{k-1} \frac{p(\mathbf{Y}_k|\mathbf{X}^n_k)}{\prod_{l=1}^{L} b^n_{l,k}}$

The more localized ambiguity function [1] a waveform has in the delay-Doppler space, the more accurate information it provides on the range and range rate of a given target with respect to the radar system. Therefore, carefully selecting the transmitted waveform is imperative to the design of an accurate radar system.

In this section, we elaborate on the fundamental concepts traditionally used in radar target tracking. These are concepts concerning the nature of the transmitted radar signal, the processing of the return signal and information extraction from the return signal.

3.5.1 RADAR SIGNAL PROCESSING

Different types of waveforms are used in radar. These are employed to achieve pulse compression, which increases range resolution, while keeping a large duty factor (the ratio of the pulse duration to the pulse period) [43]. A large duty factor allows for more transmit waveform energy, which improves the SNR. Pulse compression is accomplished with matched filtering; the output of a carefully selected waveform from the matched filter is a signal compressed in time [43]. Furthermore, we are interested in waveforms having concentrated ambiguity functions that lie in the delay-Doppler domain in order to provide good resolution in both range and range rate. In this section, we explain how the return signal is processed by a radar system and how information on range and range rate is obtained.

A radar system collects information regarding the range and range rate of a target relative to the radar sensors by transmitting signal pulses and processing the return reflected by a target of interest. The return signal bears information on the range of a target $r_{l,k}$ relative to a sensor $u = 1, \ldots, U$ of the radar system in the form of a time delay $\tau_{u,k}$ relative to the transmitted signal as $r_{u,k} = \frac{c\tau_{u,k}}{2}$, where c is the velocity of propagation of the signals. Moreover, a radar system bears information and on the range rate of a target $\dot{r}_{u,k}$ in the form of a Doppler shift $\nu_{u,k}$ as $\dot{r}_{u,k} = -\frac{c\nu_{u,k}}{2f_c}$. In the expression for range rate, f_c is the carrier frequency.

We assume that at every time step k a radar waveform $s(m)$, $m = 0, \ldots, M-1$, where M is the total number of samples of the waveform, is transmitted from U sensors. The return signals that are reflected from the target with a delay $\tau_{u,k}$ and Doppler $\nu_{u,k}$ arriving at the sensors are given by:

$$d_{u,k}(m) = A_k s(m - \tau_{u,k}) e^{j2\pi m \nu_{u,k}/M} + v(m) \tag{3.12}$$

for $m = 0, \ldots, M-1$, where A_k is a sum of random complex returns from many different target scatterers, according to the Swerling I model [44]. Therefore, A_k is assumed to be zero-mean, complex Gaussian with known variance $2\sigma_A^2$ and $v(m)$ is zero-mean complex Gaussian noise with variance $2N_0$. The return signal when no target is present is given by

$$d_{u,k}(m) = v(m), \ m = 0, \ldots, M-1 .$$

The return signal is passed through matched filters located at different time $\bar{\tau}$ and Doppler $\bar{\nu}$ shift locations with output the matched filter statistic $\mathbf{y}_{\bar{\tau},\bar{\nu},u,k} = |\tilde{\mathbf{y}}_{\bar{\tau},\bar{\nu},u,k}|^2$ where

$$\tilde{\mathbf{y}}_{\bar{\tau},\bar{\nu},u,k} = \sum_{m=0}^{M_d-1} d_{u,k}(m) s^*(m - \bar{\tau}) e^{-j2\pi m \bar{\nu}/M}$$

which using (3.12) and (2.2) becomes

$$\tilde{\mathbf{y}}_{\bar{\tau},\bar{\nu},u,k} = A_k E_s \mathrm{AF}_s(\bar{\tau} - \tau_{u,k}, \nu_{u,k} - \bar{\nu}) + \sum_{m=0}^{M_d-1} v_{u,k}(m) s^*_{\bar{\tau},\bar{\nu},u,k}(m) \tag{3.13}$$

with $M_d > M$ chosen to be large enough to accommodate a maximum delay of a signal reflected from a target. Specifically, $\tilde{\mathbf{y}}_{\bar{\tau},\bar{\nu},u,k}$ is in the absence of a target is zero mean, complex Gaussian with variance $\sigma_0^2 = 2N_0 E_s$, where E_s is the energy of the transmitted signal. In the presence of a target, it is zero mean, complex Gaussian with variance $\sigma_1^2 = 2\sigma_A^2 E_s^2 \mathcal{A}_s(\bar{\tau} - \tau_{u,k}, \nu_{u,k} - \bar{\nu}) + 2N_0 E_s$ where $\mathcal{A}_s(\bar{\tau} - \tau_{u,k}, \nu_{u,k} - \bar{\nu}) = |\mathrm{AF}_s(\bar{\tau} - \tau_{u,k}, \nu_{u,k} - \bar{\nu})|^2$ is the AF surface. The proof for σ_0^2 and σ_1^2 is provided below while a similar analysis for the continuous time case is given in [2]. The variance of $\tilde{\mathbf{y}}_{\bar{\tau},\bar{\nu},u,k}$ when the target is present is given by

$$\sigma_1^2 = E[\tilde{\mathbf{y}}_{\bar{\tau},\bar{\nu},u,k} \tilde{\mathbf{y}}^*_{\bar{\tau},\bar{\nu},u,k}]$$

which using (3.13) and the independence of A_k and $v_{u,k}(m)$ becomes

$$\begin{aligned}\sigma_1^2 = {} & E[(A_k E_s \mathrm{AF}_s(\bar{\tau} - \tau_{u,k}, \nu_{u,k} - \bar{\nu}) A_k^* E_s \mathrm{AF}_s^*(\bar{\tau} - \tau_{u,k}, \nu_{u,k} - \bar{\nu})] \\ & + E[(\sum_{m=0}^{M_d-1} v_{u,k}(m) s^*_{\bar{\tau},\bar{\nu},u,k}(m))(\sum_{m=0}^{M_d-1} v^*_{u,k}(m) s_{\bar{\tau},\bar{\nu},u,k}(m))]\end{aligned}$$

which further using the independence of $v_{u,k}(m)$ and $v_{u,k}(m')$, $m \neq m'$

$$\sigma_1^2 = E[A_k A_k^*] E_s^2 \mathrm{AF}_s(\bar{\tau} - \tau_{u,k}, \nu_{u,k} - \bar{\nu}) \mathrm{AF}_s^*(\bar{\tau} - \tau_{u,k}, \nu_{u,k} - \bar{\nu}) + \sum_{m=0}^{M_d-1} E[v_{u,k}(m) v_{u,k}^*(m)] s_{\bar{\tau},\bar{\nu},u,k}^*(m) s_{\bar{\tau},\bar{\nu},u,k}(m)$$

where using $2\sigma_A^2 = E[(A_k A_k^*)]$, $2N_0 = E[v_{u,k}(m)v_{u,k}^*(m)]$, $E_s = \sum_{m=0}^{M_d-1} s_{\bar{\tau},\bar{\nu},u,k}^*(m) s_{\bar{\tau},\bar{\nu},u,k}(m)$, and $\mathcal{A}_s(\bar{\tau} - \tau_{u,k}, \nu_{u,k} - \bar{\nu}) = |\mathrm{AF}_s(\bar{\tau} - \tau_{u,k}, \nu_{u,k} - \bar{\nu})|^2$:

$$\sigma_1^2 = 2\sigma_A^2 E_s^2 \mathcal{A}_s(\bar{\tau} - \tau_{u,k}, \nu_{u,k} - \bar{\nu}) + 2N_0 E_s \, .$$

Similarly, if no target is present $\sigma_0^2 = 2N_0 E_s$. The matched filter statistic $\mathbf{y}_{\bar{\tau},\bar{\nu},u,k}$ is exponentially distributed. Therefore,

$$p(\mathbf{y}_{\bar{\tau},\bar{\nu},u,k}|\mathbf{x}_k) \propto \begin{cases} \frac{1}{\sigma_1^2} e^{-\frac{\mathbf{y}_{\bar{\tau},\bar{\nu},u,k}}{\sigma_1^2}} & \text{, if target present} \\ \frac{1}{\sigma_0^2} e^{-\frac{\mathbf{y}_{\bar{\tau},\bar{\nu},u,k}}{\sigma_0^2}} & \text{, if target not present.} \end{cases}$$

If we threshold the matched filter output, we have the following:

$$\bar{\mathbf{y}}_{\bar{\tau},\bar{\nu},u,k} = \begin{cases} 1, & \text{if } \mathbf{y}_{\bar{\tau},\bar{\nu},u,k} \geq \mathcal{T} \\ 0, & \text{if } \mathbf{y}_{\bar{\tau},\bar{\nu},u,k} < \mathcal{T} \end{cases}$$

obtaining a detection (1) or no detection (0). We can calculate a threshold $\mathcal{T} = -2\sigma_0^2 \ln(P_f)$, where ln is the natural logarithm, based on a probability of false alarm (P_f). The probability of detection is given by $P_d = P_f^{1/(1+\mathrm{SNR}\ \mathcal{A}_s(\bar{\tau}-\tau_{u,k},\nu_{u,k}-\bar{\nu})}$. The signal-to-noise ratio (SNR) is defined as $\mathrm{SNR} = \frac{\sigma_A^2 E_s}{N_0}$, where E_s is the energy of the transmitted signal [2].

Traditionally, the delay-Doppler space is exhaustively overlaid with tessellating cells from which we obtain detections in order to infer the location of a target in the range-range rate space. This is demonstrated in Figure 3.1. The estimate of a target's position is based on averaging the grid locations, where a detection is obtained. The authors of [2, 3] use the thresholded ambiguity function shape and orientation called the "ambiguity function primitive" to create these tessellating shaped cells. This is because, excluding the effect of noise and the scatterers, the matched filter statistic is proportional to the ambiguity function. However, a tessellating shaped cell does not exactly match the shape of the thresholded ambiguity function. Therefore, increased measurement errors follow from this method. Specifically, the probability of detection is taken as the average probability of detection over the parallelogram and not just the primitive. Therefore, the area between the boundary of the primitive and the tessellating shape introduces errors. The concepts of the ambiguity function primitive and the tessellating shape are demonstrated graphically in Figure 3.2. The particle

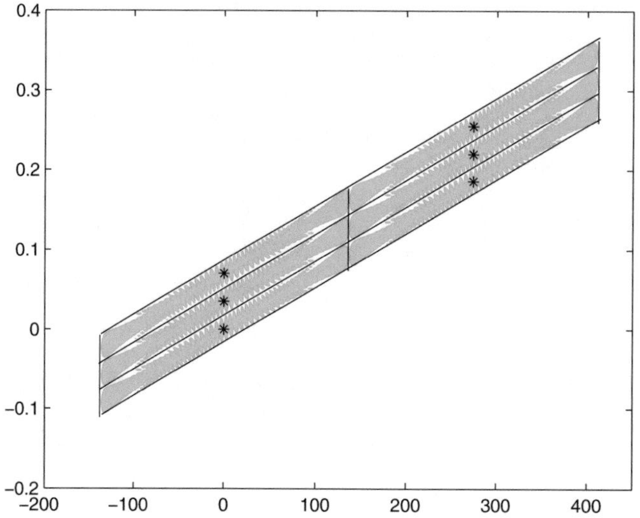

Figure 3.1: The delay-Doppler plane is overlaid with tessellating shaped resolution cells approximated by ambiguity function primitives.

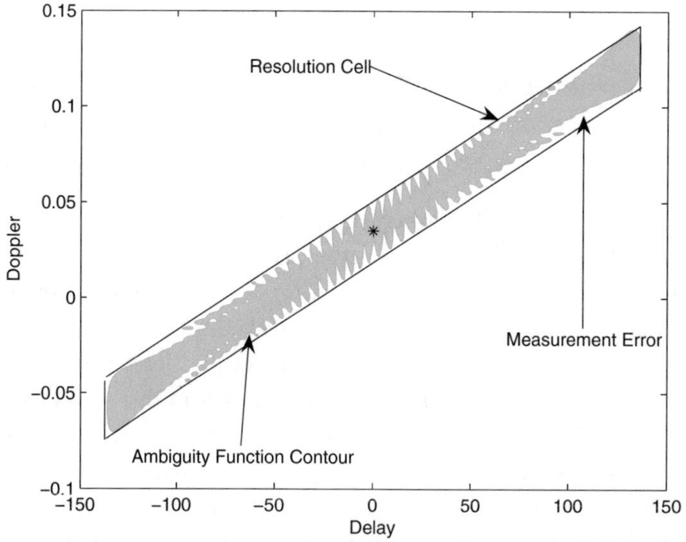

Figure 3.2: Ambiguity function primitive of an LFM waveform and the tessellating shaped resolution contour. An error in measurements is introduced by the area that lies between the boundaries of the primitive and the parallelogram.

filtering methods described in Chapters 4 and 5 avoid the need to form tessellating shaped cells and the exhaustive evaluation of measurements at grid locations across the entire delay-Doppler plane. Therefore, a particle filtering approach reduces approximation errors and the computational expense at the matched filter stage of the receiver.

CHAPTER 4

Single Target tracking with LFM and CAZAC Sequences

In this chapter, we describe a sampling importance resampling particle filter (SIRPF) and a likelihood particle filter (LPF) [4] for the radar tracking problem. The SIRPF commonly uses the prior density as the importance density. However, when the likelihood is much more concentrated than the prior, samples proposed from the prior will be spread and, thus, will receive low weights when weighted with the likelihood. Therefore, the LPF is employed that uses the likelihood as the importance density. In the case of the LFM waveform, the likelihood is broad due to the spread of its ambiguity function. Therefore, the SIRPF method can be used. In the case of the Björck CAZAC, however, the likelihood is very peaked as compared with the prior and the LPF is employed. The LPF in this setting is more computationally expensive than the SIRPF. However, the use of the LPF with the Björck CAZAC results in significant gains in tracking performance as compared to using the SIRPF with an LFM waveform as we demonstrate in Section 4.4.

4.1 MOTION MODEL

We start by describing the motion model used by both algorithms. We consider a single target moving in a two-dimensional plane. The dynamics of the target are modeled using a nearly constant velocity motion model in Cartesian coordinates. Specifically, the state vector for the target at time step k, $k = 1, \ldots, K$, is given by: $\mathbf{x}_k = [x_k \; \dot{x}_k \; y_k \; \dot{y}_k]^T$, where x_k, y_k are the positions in the x and y coordinates, and $\dot{x}_k$, $\dot{y}_k$ are the corresponding velocities. The motion is formulated as:

$$\mathbf{x}_k = \mathbf{F}\mathbf{x}_{k-1} + \mathbf{Q}\mathbf{v}_{k-1}, \tag{4.1}$$

where

$$\mathbf{F} = \begin{bmatrix} 1 & \delta t & 0 & 0 \\ 0 & 1 & 0 & 0 \\ 0 & 0 & 1 & \delta t \\ 0 & 0 & 0 & 1 \end{bmatrix}$$

and δt is the time difference between state transitions. The matrix $\mathbf{Q}$ is the process noise covariance matrix, and $\mathbf{v}_k$ denotes a zero-mean, unit variance Gaussian process that models errors in velocity, possibly due to unknown acceleration; in this work, we restrict $\mathbf{Q}$ to be diagonal. The model in (4.1) can be used to determine the kinematic prior distribution $p(\mathbf{x}_k|\mathbf{x}_{k-1})$. The measurement model was described in Section 3.5.1.

4.2 SIR PARTICLE FILTER

Every particle is a hypothesis on the state $\mathbf{x}_k$ of a target denoted by $\mathbf{x}_k^n$ where $n = 1, \ldots, N$ is the particle index and N is the total number of particles used. Having a proposed state $\mathbf{x}_{k-1}^n$ at time $k-1$ for each particle, we independently propagate each particle $n = 1, \ldots, N$ to time step k using: $\mathbf{x}_k^n = \mathbf{F}\mathbf{x}_{k-1}^n + \mathbf{Q}\mathbf{v}_k$, which amounts to sampling from a Gaussian distribution $q(\mathbf{x}_k|\mathbf{x}_{k-1}^n)$ with mean $\mathbf{F}\mathbf{x}_{k-1}^n$ and covariance matrix $\mathbf{Q}$, called the importance density [4]. The generated states $\mathbf{x}_k^n$ correspond to range and range rate

$$r_{u,k}^n = \sqrt{(\chi_u - x_k^n)^2 + (\psi_u - y_k^n)^2}$$

$$\dot{r}_{u,k}^n = (\dot{x}_k^n(x_k^n - \chi_u) + \dot{y}_k^n(y_k^n - \psi_u))/r_{u,k}^n$$

which in turn correspond to delay and Doppler shift indices $\tau_{u,k}^n = \text{round}(\frac{2\dot{r}_{u,k}^n/c}{\Delta\tau})$ and $\nu_{u,k}^n = \text{round}(\frac{-2f_c\dot{r}_{u,k}^n/c}{\Delta\nu})$, respectively. Here, χ_u and ψ_u are the x and y locations of the u sensor. The output of the matched filter at a proposed delay-Doppler location by a particle n and with respect to a sensor u is:

$$\mathbf{y}_{u,k}^n = |\sum_{m=0}^{M_d-1} d_{u,k}(m)s^*(m - \tau_{u,k}^n)e^{-j2\pi m\nu_{u,k}^n/M}|^2 .$$

The likelihood function for each particle $n' = 1, \ldots, N$ is given by

$$p^{n'}(\{\{\mathbf{y}_{u,k}^n\}_{n=1}^N\}_{u=1}^U|\mathbf{x}_k^1, \ldots, \mathbf{x}_k^{n'}, \ldots, \mathbf{x}_k^N) .$$

Here, the hypothesis of particle n' is that the state equals $\mathbf{x}_k^{n'}$, while $\{\{\mathbf{y}_{u,k}^n\}_{n=1}^N\}_{u=1}^U$ are the measurements obtained from matched filters at the delay-Doppler location defined by the particle proposed target state vectors $\{\mathbf{x}_k^n\}_{n=1}^N$. This likelihood appears in the weight equation of the particle described later in this section. However, this likelihood is a multivariate exponential distribution [45] that grows in dimensionality as N increases. We assume that the measurements $\{\{\mathbf{y}_{u,k}^n\}_{n=1}^N\}_{u=1}^U$ are independent and demonstrate the validity of this assumption in Appendix A. Assuming the measurements to be independent simplifies the likelihood for each particle to

$$\prod_{n=1,n\neq n'}^N \prod_{u=1}^U p_0^n(\mathbf{y}_{u,k}^n|\mathbf{x}_k^n)p_1^{n'}(\mathbf{y}_{u,k}^{n'}|\mathbf{x}_k^{n'}) ,$$

where $p_1^n(\mathbf{y}_{u,k}^n|\mathbf{x}_k^n)$ denotes the likelihood that a target exists at $\mathbf{x}_k^n$, and $p_0^n(\mathbf{y}_{u,k}^n|\mathbf{x}_k^n)$ denotes the likelihood that a target does not exist at $\mathbf{x}_k^n$. In Appendix A, we also show that the covariance between two measurements $\mathbf{y}_{u,k}^{n'}, \mathbf{y}_{u,k}^n$ with $n' \neq n$, depends on the closeness of the two filters relative to the ambiguity function spread. Therefore, the measurement independence assumption holds better for the Björck CAZAC and than for the LFM. Continuing with the likelihood:

$$p_1^n(\mathbf{y}_{u,k}^n|\mathbf{x}_k^n) \propto \frac{1}{\sigma_1^2} e^{-\frac{y_{u,k}^n}{\sigma_1^2}} \quad \text{if target present}$$

$$p_0^n(\mathbf{y}_{u,k}^n|\mathbf{x}_k^n) \propto \frac{1}{\sigma_0^2} e^{-\frac{y_{u,k}^n}{\sigma_0^2}} \quad \text{if target not present.}$$

Since the value of $\sigma_1^2 = 2\sigma_A^2 E_s^2 \mathcal{A}_s(\bar{\tau} - \tau_{u,k}, \nu_{u,k} - \bar{\nu}) + 2N_0 E_s$ is not known as the true target location $\tau_{u,k}, \nu_{u,k}$ is not known, thresholding of the matched filter output is required:

$$\bar{\mathbf{y}}_{u,k}^n(k) = \begin{cases} 1, & \text{if } \mathbf{y}_{u,k}^n \geq \mathcal{T} \\ 0, & \text{if } \mathbf{y}_{u,k}^n < \mathcal{T}. \end{cases}$$

The probability of detection is given by: $P_d = P_f^{1/(1+\text{SNR}\ \mathcal{A}_s(\tau_{u,k}^n - \tau_{u,k}, \nu_{u,k} - \nu_{u,k}^n))}$ and is calculated as an average over different values of $\tau_{u,k}^n - \tau_{u,k}, \nu_{u,k} - \nu_{u,k}^n$ [2].

Following the proposal, each particle is weighted as:

$$w_k^n \propto w_{k-1}^n \frac{\prod_{n=1,n\neq n'}^{N} \prod_{u=1}^{U} p_0^n(\bar{\mathbf{y}}_{u,k}^n|\mathbf{x}_k^n) p_1^{n'}(\bar{\mathbf{y}}_{u,k}^{n'}|\mathbf{x}_k^{n'}) p(\mathbf{x}_k|\mathbf{x}_{k-1}^n)}{q(\mathbf{x}_k|\mathbf{x}_{k-1}^n)}$$

which, with $q(\mathbf{x}_k|\mathbf{x}_{k-1}^n) = p(\mathbf{x}_k|\mathbf{x}_{k-1}^n)$, simplifies to

$$w_k^{n'} \propto w_{k-1}^{n'} \prod_{n=1}^{N} \prod_{u=1}^{U} p_0^n(\bar{\mathbf{y}}_{u,k}^n|\mathbf{x}_k^n) p_1^{n'}(\bar{\mathbf{y}}_{u,k}^{n'}|\mathbf{x}_k^{n'}) \,.$$

Further modifying the weights by dividing the left-hand side by the constant

$$\prod_{n=1}^{N} \prod_{u=1}^{U} p_0^n(\bar{\mathbf{y}}_{u,k}^n|\mathbf{x}_k^n) \,,$$

the weights for a particle is given by the simple expression

$$w_k^{n'} \propto w_{k-1}^{n'} \prod_{u=1}^{U} \frac{p_1^{n'}(\bar{\mathbf{y}}_{u,k}^{n'}|\mathbf{x}_k^{n'})}{p_0^{n'}(\bar{\mathbf{y}}_{u,k}^{n'}|\mathbf{x}_k^{n'})} \,.$$

Each particle then, represents a hypothesis on the delay and Doppler shift of the return signal due to the state of the target. A matched filter can, therefore, be placed at each of the proposed

Table 4.1: SIR Particle Filter [4]

- For each particle $n = 1, \ldots, N$
  - Sample
    $\mathbf{x}_k^n \sim q(\mathbf{x}_k|\mathbf{x}_{k-1}^n)$
  - For each sensor $u = 1, \ldots, U$
    * $r_{u,k}^n = \sqrt{(\chi_u - x_k^n)^2 - (\psi_u - y_k^n)^2}$
    * $\dot{r}_{u,k}^n = (\dot{x}_k^n(x_k^n - \chi_u) + \dot{y}_k^n(y_k^n - \psi_u))r_{u,k}^n$
    * $\tau_{u,k}^n = \text{round}(\frac{2\dot{r}_{u,k}^n/c}{\Delta\tau})$
    * $\nu_{u,k}^n = \text{round}(\frac{-2f_c\dot{r}_{u,k}^n/c}{\Delta\nu})$
    * $\mathbf{y}_{u,k}^n = |\sum_{m=0}^{M_d-1} d_{u,k}(m)s^*(m - \tau_{u,k}^n)e^{-j2\pi m\nu_{u,k}^n/M}|^2$
    * Obtain $\bar{\mathbf{y}}_{u,k}^n$ by thresholding $\mathbf{y}_{u,k}^n$
  - $w_k^n \propto w_{k-1}^n \prod_{u=1}^{U} \frac{p_1^n(\bar{\mathbf{y}}_{u,k}^n|\mathbf{x}_k^n)}{p_0^n(\bar{\mathbf{y}}_{u,k}^n|\mathbf{x}_k^n)}$

delay and Doppler shifts proposed by the particles. This deviates from the traditional practice of evaluating matched filters exhaustively at regular intervals across the delay-Doppler space. In the particle filter case, the matched filters are not spread evenly in the delay-Doppler space, but they are proposed at locations where a target is believed to exist at time step k. This also removes the need to design a tessellating shape to approximate probability of detection contours that introduces measurement errors. The algorithm is outlined in Table 4.1. A sample code for the SIRPF is provided in Appendix C.

4.3 LIKELIHOOD PF

In the Björck CAZAC case, the likelihood is very concentrated. Although this is good for measurement accuracy, the SIRPF will not work in this case as particles sampled from the broadly spread prior will not satisfy the likelihood easily. With the LPF, we sample values from the likelihood as it is more concentrated than the prior in the Björck CAZAC case. To achieve this, we evaluate the likelihood values at discrete bins of the delay-Doppler space of size T and $\Delta\nu$. We create a histogram from these values and sample from it. To avoid, however, the exhaustive evaluation of the likelihood in the entire delay-Doppler space, we narrow our selection of bins to those that fall within the prior. Therefore, for each particle, we select a region within which a proposal from the prior would fall with probability of almost 1.

We begin by propagating the $x - y$ coordinate components of the state space without the addition of noise. This provides the mean of the Gaussian distribution from which we would sample from in the SIRPF case:

$$[\check{x}_k^n, \check{y}_k^n] = [x_{k-1}^n + \dot{x}_k^n \tau, y_{k-1}^n + \dot{x}_k^n \tau]. \tag{4.2}$$

From this, it follows:

$$\check{r}_{u,k}^n = \sqrt{(\chi_u - \check{x}_k^n)^2 - (\psi_u - \check{y}_k^n)^2}. \tag{4.3}$$

If we assume for simplicity that $\sigma_x = \sigma_y$, then with probability of almost 1, the proposed particle will fall within $3\sigma_x$ from $\check{x}_k^n$ and similarly from $\check{y}_k^n$. Further assuming that the target is at angle $\pi/2$ with the sensor, the maximum and minimum possible sampled x and y coordinates would yield the maximum a minimum range:

$$[r_{u,k,min}^n, r_{u,k,max}^n] = [\check{r}_{u,k}^n - 3\sqrt{2}\sigma_x, \check{r}_{u,k}^n + 3\sqrt{2}\sigma_x] \tag{4.4}$$

which implies minimum and maximum index values for the delay, and with a similar procedure for the Doppler:

$$[\tau_{u,k,min}^n, \tau_{u,k,max}^n] = [\text{round}(\frac{2r_{u,k,min}^n/c}{T}), \text{round}(\frac{2r_{u,k,max}^n/c}{T})] \tag{4.5}$$

$$[\dot{r}_{u,k,min}^n, \dot{r}_{u,k,max}^n] = [\dot{r}_{u,k}^n - 3\sqrt{2}\sigma_{\dot{x}}, \dot{r}_{u,k}^n + 3\sqrt{2}\sigma_{\dot{x}}] \tag{4.6}$$

$$[\nu_{u,k,min}^n, \nu_{u,k,max}^n] = [\text{round}(\frac{-2f_c\dot{r}_{u,k,min}^n/c}{\Delta\nu}), \text{round}(\frac{-2f_c\dot{r}_{u,k,max}^n/c}{\Delta\nu})] \tag{4.7}$$

We form sets of indices for delay and Doppler between the minimum and maximum allowable values as:

$$\tau_{u,k}^n = \{\tau_{u,k,min}^n + i_\tau\}_{i_\tau=0}^{I_\tau^n}, \nu_{u,k}^n = \{\nu_{u,k,min}^n + i_\nu\}_{i_\nu=0}^{I_\nu^n}$$

where $I_\tau^n = \tau_{u,k,max}^n - \tau_{u,k,min}^n$, $I_\nu^n = \nu_{u,k,max}^n - \nu_{u,k,min}^n$. Evaluating the matched filter output at each of these values we have the following:

$$\mathbf{y}_{i_\tau,i_\nu,u,k}^n = |\sum_{m=0}^{M_d-1} d_{u,k}(m)s^*(m - \tau_{i_\tau,u,k}^n)e^{-j2\pi m\nu_{i_\nu,u,k}^n/M}|^2 .$$

Furthermore, the likelihood using the independence assumption is given:

$$p_1^n(\mathbf{y}_{i_\tau,i_\nu,u,k}^n|\mathbf{x}_k^n) \propto \frac{1}{\sigma_1^2} e^{-\frac{\mathbf{y}_{i_\tau,i_\nu,u,k}^n}{\sigma_1^2}} \quad \text{if target present}$$

$$p_0^n(\mathbf{y}_{i_\tau,i_\nu,u,k}^n|\mathbf{x}_k^n) \propto \frac{1}{\sigma_0^2} e^{-\frac{\mathbf{y}_{i_\tau,i_\nu,u,k}^n}{\sigma_0^2}} \quad \text{if target not present.}$$

Since the assumption is that a target is within the resolution cell with index $i_\tau, i_\nu, \tau_{u,k}^n - \tau_{u,k} = 0$ and $\nu_{u,k} - \nu_{u,k}^n = 0$. Therefore, in the expression of $\sigma_1^2 = \sigma_A^2 E_s^2 \mathcal{A}_s(\bar{\tau} - \tau_{u,k}, \nu_{u,k} - \bar{\nu}) + 2N_0 E_s$, we substitute $\mathcal{A}_s(\tau_{u,k}^n - \tau_{u,k}, \nu_{u,k} - \nu_{u,k}^n) = \mathcal{A}_s(0,0) = 1$. Unlike the case of the LFM, in the case of the Björck CAZAC, σ_1^2 is known. This allows us to use unthresholded measurements in the case of the Björck CAZAC for greater measurement accuracy.
After calculating the necessary likelihood values, we sample from it using the following process. Using the same assumptions for measurement independence as in the case of the SIRPF, we calculate the likelihood ratio as

$$\beta_{i_\tau,i_\nu,u,k}^n = \frac{p_1^n(\mathbf{y}_{i_\tau,i_\nu,u,k}^n|\mathbf{x}_k^n)}{p_0^n(\mathbf{y}_{i_\tau,i_\nu,u,k}^n|\mathbf{x}_k^n)}. \tag{4.8}$$

We normalize $\{\{\beta_{i_\tau,i_\nu,u,k}^n\}_{i_\tau=0}^{I_\tau^n}\}_{i_\nu=0}^{I_\nu^n}$. The normalization factor is

$$B_{u,k}^n = \sum_{i_\tau=0}^{I_\tau^n} \sum_{i_\nu=0}^{I_\nu^n} \beta_{i_\tau,i_\nu,u,k}^n \tag{4.9}$$

And the normalized distribution is:

$$b_{i_\tau,i_\nu,u,k}^n = \frac{\beta_{i_\tau,i_\nu,u,k}^n}{B_{u,k}^n}. \tag{4.10}$$

For each particle n, we sample one set of indices $\tilde{k}_\tau, \tilde{k}_\nu \sim \{\{b_{i_\tau,i_\nu,u,k}^n\}_{i_\tau=0}^{I_\tau^n}\}_{i_\nu=0}^{I_\nu^n}$. The resulting sampled range and range rate and the bias are, respectively, $\tilde{r}_{u,k}^n = c\tau_{\tilde{k}_\tau}^n T/2$, $\tilde{\dot{r}}_{u,k}^n = c\nu_{\tilde{k}_\nu}^n \Delta\nu/(-2f_c)$, $\tilde{b}_{u,k}^n = b_{\tilde{k}_\tau,\tilde{k}_\nu,u,k}^n$.

Although, the use of a CAZAC offers accuracy in the range and range rate, this does not imply accuracy in the location and velocity in the $x-y$ plane. However, with the use of two sensors and the intersections in the $x-y$ plane resulting from two range and range rate measurements, a high resolution in the $x-y$ plane results.

Continuing with the method description, the values of $\tilde{r}_{u,k}^n$ and $\tilde{\dot{r}}_{u,k}^n$ yield proposed state values for location and velocity in the $x-y$ coordinates $\mathbf{x}_k^n$. Thus, the weight equation of a particle, using likelihood ratios becomes

Table 4.2: Likelihood Particle Filter [4]

- For each particle $n = 1, \ldots, N$
  - $\check{\mathbf{x}}_k^n = \mathbf{F}\mathbf{x}_{k-1}^n$
  - For each sensor $u = 1, \ldots, U$
    - $\check{r}_{u,k}^n = \sqrt{(\chi_u - \check{x}_k^n)^2 + (\psi_u - \check{y}_k^n)^2}$
    - $\tau_{u,k}^n = \{\tau_{u,k,min}^n + i_\tau\}_{i_\tau=0}^{I_\tau^n}$
      $\nu_{u,k}^n = \{\nu_{u,k,max}^n + i_\nu\}_{i_\nu=0}^{I_\nu^n}$
      using (4.2), (5.5), (4.4), (4.5), (4.6), (4.7).
    - for $i_\tau = 0, \ldots, I_\tau^n$
    - for $i_\nu = 0, \ldots, I_\nu^n$
      - $\mathbf{y}_{i_\tau,i_\nu,u,k}^n(k) = |\sum_{m=0}^{M_d-1} d_{u,k}(m)$
        $s^*(m - \tau_{i_\tau,u,k}^n)e^{-j2\pi m\nu_{i_\nu,u,k}^n/M}|^2$
      - $\beta_{\tau_{i_\tau},\nu_{i_\nu},u,k}^n = \frac{p_1^n(\mathbf{y}_{i_\tau,i_\nu,u,k}^n|\mathbf{x}_k^n)}{p_0^n(\mathbf{y}_{i_\tau,i_\nu,u,k}^n|\mathbf{x}_k^n)}.$
    - Sample $\tilde{k}_\tau, \tilde{k}_\nu \sim \{\{b_{i_\tau,i_\nu,u,k}^n\}_{i_\tau=0}^{I_\tau^n}\}_{i_\nu=0}^{I_\nu^n}$
    - $\tilde{r}_{u,k}^n = c\tau_{\tilde{k}_\tau}^n T/2,\ \tilde{\dot{r}}_{u,k}^n = c\nu_{\tilde{k}_\nu}^n \Delta\nu/(-2f_c)$
    - $\tilde{b}_{u,k}^n = b_{\tilde{k}_\tau,\tilde{k}_\nu,u,k}^n$
  - Calculate $\mathbf{x}_k^n$ based on $\tilde{r}_{u,k}^n$ and $\tilde{\dot{r}}_{u,k}^n$
  - $w_k^n \propto w_{k-1}^n p(\mathbf{x}_k^n|\mathbf{x}_{k-1}^n) \prod_{u=1}^U B_{u,k}^n$

$$w_k^n \propto w_{k-1}^n \frac{\prod_{u=1}^U p_1^n(\mathbf{y}_{i_\tau,i_\nu,u,k}^n|\mathbf{x}_k^n)\,p(\mathbf{x}_k^n|\mathbf{x}_{k-1}^n)}{\prod_{u=1}^U p_0^n(\mathbf{y}_{i_\tau,i_\nu,u,k}^n|\mathbf{x}_k^n)\prod_{u=1}^U \tilde{b}_{u,k}^n}.$$

Using equations (4.8)–(4.10), the weight equation simplifies:

$$w_k^n \propto w_{k-1}^n p(\mathbf{x}_k^n|\mathbf{x}_{k-1}^n) \prod_{u=1}^U B_{u,k}^n .$$

The algorithm is outlined in Table 4.2. A sample code for the LPF is provided in Appendix D.

4.4 SIMULATION RESULTS AND DISCUSSION

A single target moves in a two-dimensional plane in the Cartesian coordinates. The motion is completed in 199 time steps. Two sensors located at $\chi_1 = 2000$ m, $\psi_1 = -2000$ m and $\chi_2 = 4000$ m, $\psi_2 = -2000$ m collect range and range rate measurements. The trajectory of the target and the location of the sensors are shown in Figure 4.1. The target is assumed to move according to a nearly constant velocity model with covariance matrix Q =diag(60 7 60 7). The SNR varies as 10, 12, 15, 17, 20 dB. The P_f is set to 10^{-3}, which applies only in the case of the LFM waveform. Both waveforms have $M = 4999$ samples, and they are sampled at 10 MHz and frequency modulated by $f_c = 40$ GHz. The speed of propagation of the waveforms is $c = 2.997925 \times 10^8$ m/s. For the LFM waveform, we set $f_a = 0.4$ and $f_b = 0.01$.

For the simulations, we used $N = 800$ particles, and the results were averaged over 100 Monte Carlo runs. In Figure 4.2, the RMSE tracking performance is shown for different values of SNR and for both waveforms. We observe that the LPF with a Björck CAZAC clearly outperforms the SIRPF with an LFM waveform. It also produces more reliable estimates as shown by the much shorter confidence intervals.

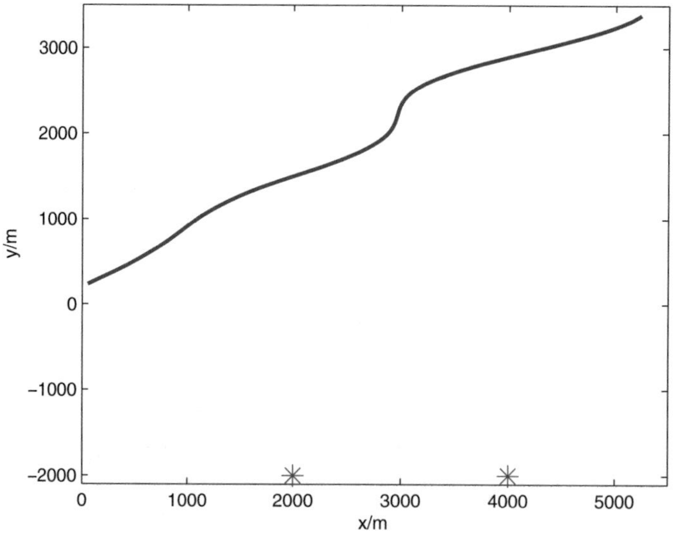

Figure 4.1: Trajectory of a single target. The two sensors are indicated by a "*".

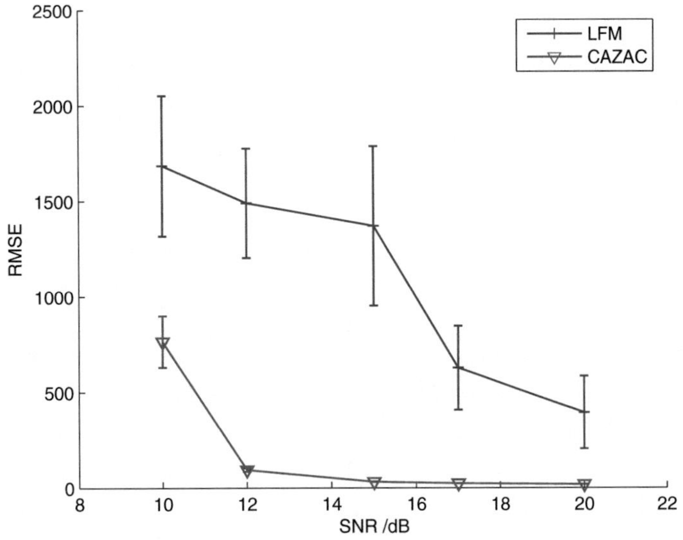

Figure 4.2: RMSE versus SNR tracking performance in a single target tracking scenario.

CHAPTER 5

Multiple Target Tracking

In this chapter, we extend the likelihood sampling concept to the case of tracking multiple targets using highly peaked AF waveforms. To this end, we combine the use of the likelihood particle filter to the independent partitions particle filter. We, moreover, demonstrate an adaptive configuration mechanism that enables the tracking of targets with significantly different cross-sectional areas.

5.1 MODEL FOR TRACKING MULTIPLE TARGETS

5.1.1 MOTION MODEL

We consider a fixed and known number of targets, denoted by L, moving in a two-dimensional plane. The dynamics of each target are modeled by a nearly constant velocity motion model [46] in Cartesian coordinates. Specifically, the state vector for the lth target, $l = 1, \ldots, L$, at time step k, $k = 1, ..., K$, is given by: $\mathbf{x}_{l,k} = [x_{l,k}\ \dot{x}_{l,k}\ y_{l,k}\ \dot{y}_{l,k}]^T$ where $x_{l,k}$, $y_{l,k}$ are the positions in the x and y coordinates, and $\dot{x}_{l,k}$, $\dot{y}_{l,k}$ are the corresponding velocities. The motion is formulated as the following:

$$\mathbf{x}_{l,k} = \mathbf{F}\mathbf{x}_{l,k-1} + \mathbf{v}_{l,k} \tag{5.1}$$

where $\mathbf{F} = [1, \Delta t, 0, 0; 0, 1, 0, 0; 0, 0, 1, \Delta t; 0, 0, 0, 1]$ and Δt is the time difference between observations. $\mathbf{v}_{l,k}$ is a white, zero mean vector Gaussian process with covariance matrix $\mathbf{Q}$ that models target deviations from constant velocity; in this work, we restrict $\mathbf{Q}$ to be diagonal. Therefore, we define $\mathbf{Q} = \text{diag}([\sigma_x^2, \sigma_y^2, \sigma_{\dot{x}}^2, \sigma_{\dot{y}}^2])$. The model in (4.1) can be used to determine the kinematic prior distribution $p(\mathbf{x}_{l,k}|\mathbf{x}_{l,k-1})$ for target l.

The multitarget state vector is expressed in terms of the state vectors of each target as

$$\mathbf{X}_k = [\mathbf{x}_{1,k}^T\ \mathbf{x}_{2,k}^T\ \ldots\ \mathbf{x}_{L,k}^T]^T$$

where $\mathbf{x}^T$ denotes the transpose of $\mathbf{x}$. Following [42], we refer to each component $\mathbf{x}_{l,k}$ of the state vector $\mathbf{X}_k$ as a partition. Since we assume that the targets move independently, the multitarget kinematic prior distribution is given by: $p(\mathbf{X}_k|\mathbf{X}_{k-1}) = \prod_{l=1}^{L} p(\mathbf{x}_{l,k}|\mathbf{x}_{l,k-1})$.

5.1.2 MEASUREMENT MODEL

A radar sensor collects information on the range and range rate of targets in the scene relative to the sensor by transmitting signal pulses and processing the returns reflected by the targets. The return signals bear information on range, in the form of a time delay, and range rate, in the form of a Doppler shift of the return signals relative to the transmitted signal. We use U sensors that operate independently, transmitting and receiving waveforms. Specifically, assuming point targets, for partition l at time step k, the range and range rate relative to sensor $u = 1, \ldots, U$ are given, respectively, by [12]:

$$r_{l,u,k} = \sqrt{(\chi_u - x_{l,k})^2 + (\psi_u - y_{l,k})^2} \, ,$$

$$\dot{r}_{l,u,k} = (\dot{x}_{l,k}(x_{l,k} - \chi_u) + \dot{y}_{l,k}(y_{l,k} - \psi_u))/r_{l,k}$$

where χ_u and ψ_u are the Cartesian coordinates of sensor u. It follows that the return signal will contain delayed and Doppler shifted replicas of the transmitted signal due to each target, with discrete delays and Doppler shifts given, respectively, by the expressions [12]: $\tau_{l,u,k} = \text{round}(\frac{2r_{l,u,k}}{cT_b})$, $\nu_{l,u,k} = \text{round}(\frac{-2f_c\dot{r}_{l,u,k}}{cMT_b})$ where round$(\cdot)$ signifies rounding to the nearest integer, c is the velocity of propagation, f_c is the carrier frequency, and T_b is the sampling period. Next, we demonstrate the processing method of the return signal and how the information on range and range rate is obtained.

5.1.3 RETURN SIGNAL AND MATCHED FILTER STATISTIC

We assume that at every time step k, a signal $s(m)$, $m = 0, \ldots, M-1$, where M is the total number of samples of the waveform, is simultaneously transmitted from each of the U sensors in different frequency bands to avoid interference. The sum of the return signals to each sensor u that are reflected from each of the L targets at the discrete delay and Doppler locations $\tau_{l,u,k}$ and $\nu_{l,u,k}$, respectively, is the following:

$$d_{u,k}(m) = \sum_{l=1}^{L} A_{l,k} s(m - \tau_{l,u,k}) e^{\frac{j2\pi m \nu_{l,u,k}}{M}} e^{\frac{j2\pi m \kappa_c}{M}} + v_{u,k}(m)$$

where κ_c is the discrete carrier frequency with which $s(m)$ is transmitted. The factor $A_{l,k}$ is a sum of random complex returns from many different target scatterers on target l, according to the Swerling I model [44]. Therefore, $A_{l,k}$ is assumed to be zero-mean, complex Gaussian with known variance $2\sigma^2_{A,l}$. In this work, it is assumed that we are tracking targets with different radar cross sections [47], therefore, targets with different strengths in their return signal. Here, the strength of the target is represented by the variance of $A_{l,k}$. In this work, we assume that the strength of the return signal depends only on the target cross section and not the distance between the sensor and the target, which is compensated by the radar by amplifying more returns that arrive later in time. The noise terms $v_{u,k}(m)$ for $u = 1, \ldots, U$ are assumed to be zero-mean complex Gaussian with variance $2N_0$ and independent for each sensor u. Here, we also define the signal-to-noise ratio (SNR) to be $\text{SNR} = \frac{\sigma^2_{A,l_w} E_s}{N_0}$ [2] where l_w is the index of the weakest target and E_s the energy of the transmitted waveform. The demodulated discrete representation of the waveform at the receiver is given:

$$d_{u,k}(m) = \sum_{l=1}^{L} A_{l,k} s(m - \tau_{l,u,k}) e^{\frac{j2\pi m \nu_{l,u,k}}{M}} + v_{u,k}(m) \,.$$

When no target is present, $d_{u,k}(m) = v_{u,k}(m)$. Even though we consider a fixed and known number of targets, the case when no target exists is necessary for the evaluation of the likelihood ratio that appears later in this section.

The return signal is filtered to match a template signal representing returns from Λ targets, at different delay and Doppler locations $\tilde{\tau}_{\lambda,u,k}$, $\tilde{\nu}_{\lambda,u,k}$, $\lambda = 1, \ldots, \Lambda$, respectively. These delay and Doppler locations are derived from the belief in target state as represented by the particles of a particle filtering approach. In this work, we assume that the number of targets is known and fixed. Therefore, we will let Λ to equal L or 1, depending whether we are proposing particles with L partitions or independently proposing a single partition. Given a set of delay and Doppler values $\{\tilde{\tau}_{\lambda,u,k}\}$ and $\{\tilde{\nu}_{\lambda,u,k}\}$, the matched filter output $\tilde{\mathbf{y}}_{u,k}$ is formed:

$$\begin{aligned}
\tilde{\mathbf{y}}_{u,k} &= \sum_{m=0}^{M_d-1} d_{u,k}(m) \sum_{\lambda=1}^{\Lambda} s^*(m - \tilde{\tau}_{\lambda,u,k}) e^{-\frac{j2\pi m \tilde{\nu}_{\lambda,u,k}}{M}} \\
&= \sum_{m=0}^{M_d-1} \sum_{l=1}^{L} \sum_{\lambda=1}^{\Lambda} (A_{l,k} s(m - \tau_{l,u,k}) e^{\frac{j2\pi m \nu_{l,u,k}}{M}} \\
&\quad + v_{u,k}(m)) s^*(m - \tilde{\tau}_{\lambda,u,k}) e^{-\frac{j2\pi m \tilde{\nu}_{\lambda,u,k}}{M}}
\end{aligned}$$

where $M_d > M$ should be large enough to accommodate a maximum delay of a signal reflected from a target. Using (2.2) the above expression becomes

$$\begin{aligned}
\tilde{\mathbf{y}}_{u,k} &= \sum_{l=1}^{L} \sum_{\lambda=1}^{\Lambda} A_{l,k} E_s \mathrm{AF}_s(\tilde{\tau}_{\lambda,u,k} - \tau_{l,u,k}, \nu_{l,u,k} - \tilde{\nu}_{\lambda,u,k}) + \\
& \sum_{m=0}^{M_d-1} v_{u,k}(m) \sum_{\lambda=1}^{\Lambda} s^*(m - \tilde{\tau}_{\lambda,u,k}) e^{-\frac{j2\pi m \tilde{\nu}_{\lambda,u,k}}{M}} \,.
\end{aligned}$$

The matched filter statistic that we will use for estimation is given by $\mathbf{y}_{u,k} = |\tilde{\mathbf{y}}_{u,k}|^2$.

5.1.4 MEASUREMENT LIKELIHOOD

We proceed to identify the distribution of $\tilde{\mathbf{y}}_{u,k}$ and its parameters which will lead to the characterization of the statistical properties of the matched filter statistic $\mathbf{y}_{u,k}$. Since $A_{l,k}$ and $v_{u,k}(m)$ are complex Gaussian with zero mean, $\tilde{\mathbf{y}}_{u,k}$ is complex Gaussian with zero mean. Below we derive variance of $\tilde{\mathbf{y}}_{u,k}$ in the presence of L targets denoted by

$$\sigma_1^2 = E[\tilde{\mathbf{y}}_{u,k}\tilde{\mathbf{y}}_{u,k}^*] \, .$$

Due to the independence of $A_{l,k}$ and $v_{u,k}(m)$, this becomes:

$$\sigma_1^2 = E_s^2 E[(\sum_{l=1}^{L}\sum_{\lambda=1}^{\Lambda} A_{l,k}\mathrm{AF}_s(\tilde{\tau}_{\lambda,u,k} - \tau_{l,u,k}, \nu_{l,u,k} - \tilde{\nu}_{\lambda,u,k}))$$

$$(\sum_{l=1}^{L}\sum_{\lambda'=1}^{\Lambda} A_{l,k}^*\mathrm{AF}_s^*(\tilde{\tau}_{\lambda',u,k} - \tau_{l,u,k}, \nu_{l,u,k} - \tilde{\nu}_{\lambda',u,k}))]$$

$$+E[(\sum_{m=0}^{M_d-1} v_{u,k}(m)\sum_{\lambda=1}^{\Lambda} s^*(m - \tilde{\tau}_{\lambda,u,k})e^{-\frac{j2\pi m\tilde{\nu}_{\lambda,u,k}}{M}})$$

$$(\sum_{m=0}^{M_d-1} v_{u,k}^*(m)\sum_{\lambda'=1}^{\Lambda} s(m - \tilde{\tau}_{\lambda',u,k})e^{\frac{j2\pi m\tilde{\nu}_{\lambda',u,k}}{M}})] \, .$$

Moreover, due to the independence of $A_{l,k}$ for each $l = 1, \ldots, L$ and $v_{u,k}(m)$ for each $m = 0, \ldots, M-1$:

$$\sigma_1^2 = E_s^2 \sum_{l=1}^{L} E[(\sum_{\lambda=1}^{\Lambda} A_{l,k}\mathrm{AF}_s(\tilde{\tau}_{\lambda,u,k} - \tau_{l,u,k}, \nu_{l,u,k} - \tilde{\nu}_{\lambda,u,k}))$$

$$(\sum_{\lambda'=1}^{\Lambda} A_{l,k}^*\mathrm{AF}_s^*(\tilde{\tau}_{\lambda',u,k} - \tau_{l,u,k}, \nu_{l,u,k} - \tilde{\nu}_{\lambda',u,k}))]$$

$$+\sum_{m=0}^{M_d-1} E[(v_{u,k}(m)\sum_{\lambda=1}^{\Lambda} s^*(m - \tilde{\tau}_{\lambda,u,k})e^{-\frac{j2\pi m\tilde{\nu}_{\lambda,u,k}}{M}})(v_{u,k}^*(m)\sum_{\lambda'=1}^{\Lambda} s(m - \tilde{\tau}_{\lambda',u,k})e^{\frac{j2\pi m\tilde{\nu}_{\lambda',u,k}}{M}})]$$

$$= E_s^2 \sum_{l=1}^{L} E[A_{l,k}A_{l,k}^*](\sum_{\lambda=1}^{\Lambda} \mathrm{AF}_s(\tilde{\tau}_{\lambda,u,k} - \tau_{l,u,k}, \nu_{l,u,k} - \tilde{\nu}_{\lambda,u,k}))$$

$$(\sum_{\lambda'=1}^{\Lambda} \mathrm{AF}_s^*(\tilde{\tau}_{\lambda',u,k} - \tau_{l,u,k}, \nu_{l,u,k} - \tilde{\nu}_{\lambda',u,k}))$$

$$+\sum_{m=0}^{M_d-1} E[v_{u,k}(m)v_{u,k}^*(m)](\sum_{\lambda=1}^{\Lambda} s^*(m - \tilde{\tau}_{\lambda,u,k})e^{-\frac{j2\pi m\tilde{\nu}_{\lambda,u,k}}{M}})(\sum_{\lambda'=1}^{\Lambda} s(m - \tilde{\tau}_{\lambda',u,k})e^{\frac{j2\pi m\tilde{\nu}_{\lambda',u,k}}{M}})$$

$$= 2E_s^2 \sum_{l=1}^{L} \sigma_{A,l}^2 \sum_{\lambda=1}^{\Lambda} \sum_{\lambda'=1}^{\Lambda} \mathrm{AF}_s(\tilde{\tau}_{\lambda,u,k} - \tau_{l,u,k}, \nu_{l,u,k} - \tilde{\nu}_{\lambda,u,k}) \mathrm{AF}_s^*(\tilde{\tau}_{\lambda',u,k} - \tau_{l,u,k}, \nu_{l,u,k} - \tilde{\nu}_{\lambda',u,k})$$

$$+2N_0 \sum_{\lambda=1}^{\Lambda} \sum_{\lambda'=1}^{\Lambda} \sum_{m=0}^{M_d-1} s^*(m - \tilde{\tau}_{\lambda,u,k}) e^{-\frac{j2\pi m\tilde{\nu}_{\lambda,u,k}}{M}} s(m - \tilde{\tau}_{\lambda',u,k}) e^{\frac{j2\pi m\tilde{\nu}_{\lambda',u,k}}{M}} \Rightarrow$$

$$\sigma_1^2 = 2E_s^2 \sum_{l=1}^{L} \sigma_{A,l}^2 \sum_{\lambda=1}^{\Lambda} \sum_{\lambda'=1}^{\Lambda} \mathrm{AF}_s(\tilde{\tau}_{\lambda,u,k} - \tau_{l,u,k}, \nu_{l,u,k} - \tilde{\nu}_{\lambda,u,k}) \mathrm{AF}_s^*(\tilde{\tau}_{\lambda',u,k} - \tau_{l,u,k}, \nu_{l,u,k} - \tilde{\nu}_{\lambda',u,k})$$

$$+2N_0 E_s \sum_{\lambda=1}^{\Lambda} \sum_{\lambda'=1}^{\Lambda} \mathrm{AF}_s(\tilde{\tau}_{\lambda,u,k} - \tilde{\tau}_{\lambda',u,k}, \tilde{\nu}_{\lambda',u,k} - \tilde{\nu}_{\lambda,u,k}) \, .$$

The variance of $\tilde{\mathbf{y}}_{u,k}$ when no targets are present, is similarly found to be $\sigma_0^2 = 2N_0 E_s \sum_{\lambda=1}^{\Lambda} \sum_{\lambda'=1}^{\Lambda} \mathrm{AF}_s(\tilde{\tau}_{\lambda,u,k} - \tilde{\tau}_{\lambda',u,k}, \tilde{\nu}_{\lambda',u,k} - \tilde{\nu}_{\lambda,u,k})$. Since $\tilde{\mathbf{y}}_{\tilde{\tau},\tilde{\nu},u,k}$ is complex Gaussian, the matched filter statistic $\mathbf{y}_{u,k} = |\tilde{\mathbf{y}}_{u,k}|^2$ is exponentially distributed, and the measurement likelihood is given by

$$\begin{aligned} p_1(\mathbf{y}_{u,k}|\mathbf{x}_k) &= \frac{1}{2\sigma_1^2} e^{-\frac{\mathbf{y}_{u,k}}{2\sigma_1^2}} \quad \text{, if } L \text{ targets are present} \\ p_0(\mathbf{y}_{u,k}|\mathbf{x}_k) &= \frac{1}{2\sigma_0^2} e^{-\frac{\mathbf{y}_{u,k}}{2\sigma_0^2}} \quad \text{, if no target is present.} \end{aligned} \tag{5.2}$$

Note that if the template signal considers only 1 partition, λ, we have $\Lambda = 1$ and the variances of $\tilde{\mathbf{y}}_{u,k}$ are given by $\sigma_{\lambda,1}^2 = 2E_s^2 \sum_{l=1}^{L} \sigma_{A,l}^2 A_s(\tilde{\tau}_{\lambda,u,k} - \tau_{l,u,k}, \nu_{l,u,k} - \tilde{\nu}_{\lambda,u,k}) + 2N_0 E_s$ and $\sigma_{\lambda,0}^2 = 2N_0 E_s$ and the single partition likelihoods are

$$\begin{aligned} p_1(\mathbf{y}_{\lambda,u,k}|\mathbf{x}_k) &= \frac{1}{2\sigma_{\lambda,1}^2} e^{-\frac{\mathbf{y}_{\lambda,u,k}}{2\sigma_{\lambda,1}^2}} \quad \text{if target } \lambda \text{ is present} \\ p_0(\mathbf{y}_{\lambda,u,k}|\mathbf{x}_k) &= \frac{1}{2\sigma_{\lambda,0}^2} e^{-\frac{\mathbf{y}_{\lambda,u,k}}{2\sigma_{\lambda,0}^2}} \quad \text{if target } \lambda \text{ is not present.} \end{aligned} \tag{5.3}$$

It is, moreover, useful in the implementation of the method described in this work to consider an approximation to the variance that reduces computational expense and allows the use of unthresholded measurements in the Björck CAZAC case. Since the sidelobes of the AF of the single and adaptively configured MCPC CAZAC waveforms are small at the locations interrogated by the particle filtering method compared to the mainlobe we approximate $\mathrm{AF}_s(\tau, \nu)$ to be 0 if $\tau \neq 0$ and $\nu \neq 0$. Moreover, we require L to be small for this approximation so that the sidelobes of the AF that we assume to be zero do not add up to a significant amount. Furthermore, using the fact that $\mathrm{AF}_s(0,0) = 1, \sigma_1^2 = 2E_s^2 L\sigma_A^2 + 2N_0 E_s L, \sigma_0^2 = 2N_0 E_s L, \sigma_{\lambda,1}^2 = 2E_s^2 L\sigma_A^2 + 2N_0 E_s$, and $\sigma_{\lambda,0}^2 = 2N_0 E_s$. Above, we let $\sigma_{A,l}^2 = \sigma_A^2$ for all l, where σ_A^2 is some nominal value that we choose, since we assume the target strength to be unknown.

5.2 INDEPENDENT PARTITION LIKELIHOOD PARTICLE FILTER ALGORITHM

5.2.1 LIKELIHOOD PARTITION SAMPLING

The highly concentrated AF of a Björck CAZAC sequence provides a highly concentrated likelihood proposal distribution. Although this is good for measurement accuracy, we need to modify the proposal process to sample particles from the likelihood instead of the kinematic prior since the former is much more localized than the latter. To achieve this, we use a likelihood based particle filter [4] where the importance density depends on the measurements rather than the kinematic prior.

In addition to the likelihood proposal, we adopt the independent partition (IP) particle filtering [42, 41] concept to efficiently propose particles. In the IP, we propose individual partitions of the multi-target state vector, each representing the state of a single target. We then combine the more accurate partition proposals into particles. The IP algorithm is an approximation to the joint multitarget probability density (JMPD) particle filter [41]; the approximation is accurate when the targets are well separated in the observation space. When targets are close in measurement space, their partitions cannot be independently proposed as described above. Due to our use of of Björck CAZAC sequences with a sharply peaked AF, measurements are well approximated as independent (see Appendix A). Our IPLPF algorithm belongs to the class of sequential partition algorithms [48]. Algorithms of this class propose partitions sequentially, whether independently or not, and then combine them into particles.

The IPLPF functions as follows. We evaluate likelihood values at discrete bins of the delay-Doppler space for each partition independently. We then create a histogram from these values and sample partition states from it. We narrow, however, our selection of bins to a region in which a partition sample from the kinematic prior would fall with probability of almost 1. This way, the number of bins required to build the histogram is reduced and the sample from the measurements is made consistent with the kinematic prior. We then evaluate partition weights by combining measurements from the different sensors and using the kinematic prior. Using the normalized partition weights we sample values for each partition independently. Next, we combine the sampled partitions into particles. The proposal of partitions is followed by the weighting of particles, estimation and particle resampling.

In order to extract useful information on the Cartesian coordinates of multiple targets from range and range rate measurements, we use multiple radar sensors. In the case of a single target [5], range information from two sensors, combined with kinematic prior knowledge was enough to produce $\chi - \psi$ coordinate estimates. To achieve this, we used geometry to find the intersection between two circles in the Cartesian coordinates with radii, the ranges of the two sensors. The two intersections of these circles provided two $\chi - \psi$ coordinate locations, one of which can be selected that agrees with the kinematic prior information. In the presence of multiple targets, however, we have multiple of these circles for each sensor and each target and multiple intersection points that do not correspond to true $\chi - \psi$ coordinate locations. Therefore, the use of three sensors helps clear the ambiguity by providing fewer intersections of three circles. In order to avoid complicated geometry, we first process the returns of two sensors, and sample Cartesian coordinate target locations using the likelihood. We then weight the sampled locations with measurements from the third sensor. Our method is for a general number of sensors equal or greater than three. In this work, we keep the number of sensors to a minimum of three.

The sampling of partitions based on the likelihood is performed in two stages. In the first stage, we utilize information from only two of the sensors in order to propose a preliminary set of partitions. This way, we avoid the complex geometry required to sample Cartesian locations from range and range rate information obtained from three or more sensors. In the second stage, we refine our selection of partitions by sampling from the preliminary set of partitions created in the first stage using information from all the sensors.

5.2.1.1 Stage 1

We begin by propagating the $\chi - \psi$ coordinate components of each partition λ of the state space without the addition of noise. We denote with λ the partition that we currently propose and with l the partition that represents the true state of the lth target. This would be the mean of the Gaussian distribution from which we would sample if we used an SIR particle filter:

$$\check{\mathbf{x}}^n_{\lambda,k} = [\check{x}^n_{\lambda,k}, \check{y}^n_{\lambda,k}, \check{\dot{x}}^n_{\lambda,k}, \check{\dot{y}}^n_{\lambda,k}]^T = \mathbf{F}\mathbf{x}^n_{\lambda,k-1}. \tag{5.4}$$

Using $\check{\mathbf{x}}^n_{\lambda,k}$, we have:

$$\check{r}^n_{\lambda,u,k} = \sqrt{(\chi_u - \check{x}^n_{\lambda,k})^2 + (\psi_u - \check{y}^n_{\lambda,k})^2}. \tag{5.5}$$

Using the method described next, we determine a region of delay-Doppler bins that could contain observations if the true state is $\check{\mathbf{x}}^n_{\lambda,k}$. This region is determined by the spread of the kinematic prior that determines the possible states of partition λ. If we assume for simplicity that the variance of the kinematic prior in the χ and ψ coordinates are equal ($\sigma_x^2 = \sigma_y^2$), then with probability of almost 1, the proposed particle will fall within $3\sigma_x$ from $\check{x}^n_{\lambda,k}$ and similarly from $\check{y}^n_{\lambda,k}$. The maximum and minimum possible sampled x and y coordinates would yield the maximum and minimum range as the following:

$$[r^n_{min,\lambda,u,k}, r^n_{max,\lambda,u,k}] = [\check{r}^n_{\lambda,u,k} - 3\sqrt{2}\sigma_x, \check{r}^n_{\lambda,u,k} + 3\sqrt{2}\sigma_x] \tag{5.6}$$

if we assume that the target is at angle $\pi/2$ with the sensor. In that case, the range would increase/decrease by the amount $3\sqrt{2}\sigma_x = \sqrt{(3\sigma_x)^2 + (3\sigma_x)^2}$. This implies minimum and maximum index values for the delay:

$$[\tau^n_{min,\lambda,u,k}, \tau^n_{max,\lambda,u,k}] = [\lfloor \frac{2r^n_{min,\lambda,u,k}}{cT} \rfloor, \lceil \frac{2r^n_{max,\lambda,u,k}}{cT} \rceil]. \tag{5.7}$$

With a similar procedure for the Doppler, we have:

$$[\dot{r}^n_{min,\lambda,u,k}, \dot{r}^n_{max,\lambda,u,k}] = [\check{\dot{r}}^n_{\lambda,u,k} - 3\sqrt{2}\sigma_{\dot{x}}, \check{\dot{r}}^n_{\lambda,u,k} + 3\sqrt{2}\sigma_{\dot{x}}], \text{ and} \tag{5.8}$$

$$[\nu^n_{min,\lambda,u,k}, \nu^n_{max,\lambda,u,k}] = [\lfloor \frac{-2f_c\dot{r}^n_{min,\lambda,u,k}}{c\Delta\nu} \rfloor, \lceil \frac{-2f_c\dot{r}^n_{max,\lambda,u,k}}{c\Delta\nu} \rceil]. \tag{5.9}$$

We form the set

$$\{\{\tau^n_{i_\tau,\lambda,u,k}, \nu^n_{i_\nu,\lambda,u,k}\}^{I^n_\tau}_{i_\tau=0}\}^{I^n_\nu}_{i_\nu=0}, \tag{5.10}$$

where $I^n_\tau = \tau^n_{max,\lambda,u,k} - \tau^n_{min,\lambda,u,k}$, $I^n_\nu = \nu^n_{max,\lambda,u,k} - \nu^n_{min,\lambda,u,k}$ containing all combinations of indices for delay and Doppler that lie within the delay and Doppler minimum and maximum values, where each index pair is given by:

$$\{\tau^n_{i_\tau,\lambda,u,k}, \nu^n_{i_\nu,\lambda,u,k}\} = \{\{\tau^n_{min,\lambda,u,k} + i_\tau, \nu^n_{min,\lambda,u,k} + i_\nu\}^{I^n_\tau}_{i_\tau=0}\}^{I^n_\nu}_{i_\nu=0}. \tag{5.11}$$

Evaluating the matched filter output (where the subscript k is omitted for simplicity) at each of these values, and for sensors $u = 1, 2$, we have:

$$\mathbf{y}^n_{i_\tau,i_\nu,\lambda,u} = |\frac{1}{M}\sum_{m=0}^{M_d-1} d_{u,k}(m)s^*(m - \tau^n_{i_\tau,\lambda,u,k})e^{-\frac{j2\pi m\nu^n_{i_\nu,\lambda,u,k}}{M}}|^2 \Rightarrow$$

$$\begin{aligned}\mathbf{y}^n_{i_\tau,i_\nu,\lambda,u} = |&\sum_{l=1}^{L} A_{l,k}E_s\text{AF}_s(\tau^n_{i_\tau,\lambda,u,k} - \tau_{l,u,k}, \nu_{l,u,k} - \nu^n_{i_\nu,\lambda,u,k})e^{-\frac{j2\pi\tau_{\lambda,u,k}\kappa_c}{M}} \\ &+ \frac{1}{M}\sum_{m=0}^{M_d-1} v_{u,k}(m)s^*(m - \tau^n_{i_\tau,\lambda,u,k})e^{-\frac{j2\pi m\nu^n_{i_\nu,\lambda,u,k}}{M}}|^2.\end{aligned} \tag{5.12}$$

Above, we have used only one delay-Doppler pair $(\tau^n_{i_\tau,\lambda,u,k}, \nu^n_{i_\nu,\lambda,u,k})$ in the template signal representing a single partition (λ). The single partition likelihood for each delay-Doppler bin $\{\tau^n_{i_\tau,\lambda,u,k}, \nu^n_{i_\nu,\lambda,u,k}\}$ is then given by the following:

$$p_1^n(\mathbf{y}_{i_\tau,i_\nu,\lambda,u}^n|\tau_{i_\tau,\lambda,u,k}^n,\nu_{i_\nu,\lambda,u,k}^n) = \frac{1}{2\sigma_{\lambda,1}^2}e^{-\frac{\mathbf{y}_{i_\tau,i_\nu,\lambda,u,k}^n}{2\sigma_{\lambda,1}^2}} \quad \text{if target } \lambda \text{ present}$$
$$p_0^n(\mathbf{y}_{i_\tau,i_\nu,\lambda,u}^n|\tau_{i_\tau,\lambda,u,k}^n,\nu_{i_\nu,\lambda,u,k}^n) = \frac{1}{2\sigma_{\lambda,0}^2}e^{-\frac{\mathbf{y}_{i_\tau,i_\nu,\lambda,u,k}^n}{2\sigma_{\lambda,0}^2}} \quad \text{if target } \lambda \text{ not present} \tag{5.13}$$

We evaluate the likelihood ratio for each delay-Doppler bin $\{\tau_{i_\tau,\lambda,u,k}^n, \nu_{i_\nu,\lambda,u,k}^n\}$ as

$$\breve{\beta}_{i_\tau,i_\nu,\lambda,u}^n = \frac{p_1^n(\mathbf{y}_{i_\tau,i_\nu,\lambda,u}^n|\tau_{i_\tau,\lambda,u,k}^n,\nu_{i_\nu,\lambda,u,k}^n)}{p_0^n(\mathbf{y}_{i_\tau,i_\nu,\lambda,u}^n|\tau_{i_\tau,\lambda,u,k}^n,\nu_{i_\nu,\lambda,u,k}^n)}. \tag{5.14}$$

We normalize $\{\{\breve{\beta}_{i_\tau,i_\nu,\lambda,u}^n\}_{i_\tau=0}^{I_\tau^n}\}_{i_\nu=0}^{I_\nu^n}$ with the normalization factor

$$\breve{B}_{\lambda,u}^n = \sum_{i_\tau=0}^{I_\tau^n}\sum_{i_\nu=0}^{I_\nu^n}\breve{\beta}_{i_\tau,i_\nu,\lambda,u}^n \tag{5.15}$$

and the normalized distribution is the following:

$$\breve{b}_{i_\tau,i_\nu,\lambda,u}^n = \frac{\breve{\beta}_{i_\tau,i_\nu,\lambda,u}^n}{\breve{B}_{\lambda,u}^n}. \tag{5.16}$$

We then sample indices $j_{i_\tau,u}^n, j_{i_\nu,u}^n \sim \{\{\breve{b}_{i_\tau,i_\nu,\lambda,u}^n\}_{i_\tau=0}^{I_\tau^n}\}_{i_\nu=0}^{I_\nu^n}$, for each particle n and each sensor $u = 1, 2$. The resulting sampled range and range rate and the bias are, respectively, $r_{i'_\tau,i'_\nu,\lambda,u}^n = \frac{c\tau_{j_{i_\tau},\lambda,u,k}^n T_b}{2}$, $\dot{r}_{i'_\tau,i'_\nu,\lambda,u}^n = -\frac{c\nu_{j_{i_\nu},\lambda,u,k}^n \Delta\nu}{2f_c}$, $b_{i'_\tau,i'_\nu,\lambda,u}^n = \breve{b}_{j_{i_\tau},j_{i_\nu},\lambda,u}^n$. The values of $r_{i'_\tau,i'_\nu,\lambda,u}^n$ and $\dot{r}_{i'_\tau,i'_\nu,\lambda,u}^n$, in turn, yield proposed state values for location and velocity in the $\chi - \psi$ coordinates $\tilde{\mathbf{x}}_{\lambda,k}^n$. This is accomplished by taking the intersection of two circles in the Cartesian plane and choosing the intersection point in Euclidian distance to the propagated $\chi - \psi$ coordinate locations for partition λ, $\check{x}_{\lambda,k}^n$, $\check{y}_{\lambda,k}^n$, i.e., the location that mostly agrees with the kinematic prior information). The above process is depicted in Figure 5.1. We note that the so far sampled partitions $\tilde{\mathbf{x}}_{\lambda,k}^n$ are based on information provided from only two sensors. Therefore, some of these partitions may be erroneous, as we explained in the introduction of this section. However, the values $\tilde{\mathbf{x}}_{\lambda,k}^n$ now allow us to easily evaluate the likelihoods for each of the three sensors in order to sample partitions in the next stage of the proposal described in Section 5.2.1.2. The information provided by the third sensor, used in the next step of the proposal process, removes partitions that have been erroneously sampled.

5.2.1.2 Stage 2

In this section, we describe the method of weighting and sampling the partitions $\{\tilde{\mathbf{x}}_{\lambda,k}^n\}_{n=1}^N$, generated from sampling from the delay-Doppler bins associated with sensors $u = 1, 2$ as described

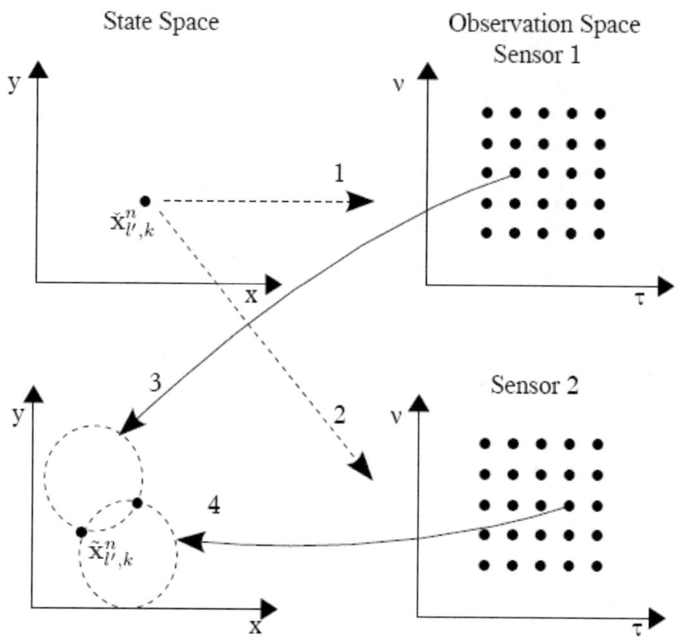

Figure 5.1: Schematic of the likelihood proposal process described in Section 5.2.1.1. We deterministically propagate forward each particle $\check{\mathbf{x}}^n_{\lambda,k}$ (left top figure, arrows 1, 2), define observation points for sensors 1 and 2 (right top and right bottom), sample one point from each observation set of each sensor (arrows 3, 4), and select one of the two states ($\tilde{\mathbf{x}}^n_{\lambda,k}$) formed by the sampled ranges in the Cartesian coordinates that agrees more with the prior (left bottom).

above. Here, we utilize the return signals transmitted by all sensors in order to refine our choice of partitions that more accurately represent the target state. The sampling method described in this section is applicable for any $U \geq 3$. Before describing the sampling process, however, we describe the complexity in calculating the partition weights and the approximation that we use to make the computations tractable. The measurements from sensors $u = 1, \ldots, U$ that are generated after passing the return signal through matched filters at locations in the delay-Doppler plane derived from proposed partitions $\{\tilde{\mathbf{x}}^n_{\lambda,k}\}_{n=1}^N$ are given by $\mathbf{y}^n_{\lambda,u,k}$, $n = 1, \ldots, N$, $u = 1, \ldots, U$ where

$$\mathbf{y}^n_{\lambda,u,k} = |\frac{1}{M} \sum_{m=0}^{M_d-1} d_{u,k}(m) s^*(m - \tilde{\tau}^n_{\lambda,u,k}) e^{-\frac{j2\pi m \tilde{\nu}^n_{\lambda,u,k}}{M}}|^2 \Rightarrow$$

$$\mathbf{y}^n_{\lambda,u,k} = |\sum_{l=1}^{L} A_{l,k} E_s \mathrm{AF}_s(\tilde{\tau}^n_{\lambda,u,k} - \tau_{l,u,k}, \nu_{l,u,k} - \tilde{\nu}^n_{\lambda,u,k}) + \frac{1}{M} \sum_{m=0}^{M_d-1} v_{u,k}(m) s^*(m - \tilde{\tau}^n_{\lambda,u,k}) e^{-\frac{j2\pi m \tilde{\nu}^n_{\lambda,u,k}}{M}}|^2 \tag{5.17}$$

with $\{\tilde{\tau}^n_{\lambda,u,k}, \tilde{\nu}^n_{\lambda,u,k}\}$ representing the delay-Doppler pair corresponding to state $\tilde{\mathbf{x}}^n_{\lambda,k}$ and the location of sensor u and $\{\tau_{l,u,k}, \nu_{l,u,k}\}$ representing the true target state $\mathbf{x}_{l,k}$. Therefore, the single partition likelihood function for each proposed partition λ of particle $n' = 1, \ldots, N$ is represented as $\prod_{u=1}^{U} p^{n'}_{\lambda}(\{\mathbf{y}^n_{\lambda,u,k}\}_{n=1}^{N} | \tilde{\mathbf{x}}^1_{\lambda,k}, \ldots, \tilde{\mathbf{x}}^{n'}_{\lambda,k}, \ldots, \tilde{\mathbf{x}}^N_{\lambda,k})$. Here, the hypothesis of particle n' and partition λ is that the partition state equals $\tilde{\mathbf{x}}^{n'}_{\lambda,k}$ and not $\tilde{\mathbf{x}}^n_{\lambda,k}$ for $n \neq n'$, while $\mathbf{y}^n_{\lambda,u,k}$, $n = 1, \ldots, N$, $u = 1, \ldots, U$ are the measurements obtained from matched filters at the delay-Doppler location defined by the particle proposed target state vectors $\tilde{\mathbf{x}}^n_{\lambda,k}$, $n = 1, \ldots, N$. However, each likelihood for sensor u is a multivariate exponential distribution [45] that grows in dimensionality as N increases. We approximate the likelihood for each partition to $\prod_{u=1}^{U} p^{n'}_1(\mathbf{y}^{n'}_{\lambda,u,k}|\tilde{\mathbf{x}}^{n'}_{\lambda,k}) \prod_{n=1,n\neq n'}^{N} p^n_0(\mathbf{y}^n_{\lambda,u,k}|\tilde{\mathbf{x}}^n_{\lambda,k})$, where $p^n_1(\mathbf{y}^n_{\lambda,u,k}|\tilde{\mathbf{x}}^n_{\lambda,k})$ denotes the likelihood that a target exists at $\tilde{\mathbf{x}}^n_{\lambda,k}$ and $p^n_0(\mathbf{y}^n_{\lambda,u,k}|\tilde{\mathbf{x}}^n_{\lambda,k})$ denotes the likelihood that a target does not exist at $\tilde{\mathbf{x}}^n_{\lambda,k}$. In Appendix A, we show that the covariance between two measurements $\mathbf{y}^{n'}_{\lambda,u,k}, \mathbf{y}^n_{\lambda,u,k}$ with $n' \neq n$, depends on the proximity of the two filters (i.e., the closeness of $\tilde{\tau}^{n'}_{\lambda,u,k}, \tilde{\nu}^{n'}_{\lambda,u,k}$ and $\tilde{\tau}^n_{\lambda,u,k}, \tilde{\nu}^n_{\lambda,u,k}$) relative to the ambiguity function spread. Therefore, the measurement independence approximation is quite reasonable for the Björck CAZAC that has a concentrated ambiguity function. Using the approximation above, the weights for partition λ of particle $n' = 1, \ldots, N$ are

$$\tilde{\beta}^{n'}_{\lambda,k} \propto \frac{\prod_{u=1}^{U} p^{n'}_1(\mathbf{y}^{n'}_{\lambda,u,k}|\tilde{\mathbf{x}}^{n'}_{\lambda,k}) \prod_{n=1,n\neq n'}^{N} p^n_0(\mathbf{y}^n_{\lambda,u,k}|\tilde{\mathbf{x}}^n_{\lambda,k})}{\prod_{u'=1}^{2} b^{n'}_{i'_\tau,i'_\nu,\lambda,u',k}} p(\tilde{\mathbf{x}}^{n'}_{\lambda,k}|\tilde{\mathbf{x}}^{n'}_{\lambda,k-1})$$

where we used the bias with which $\tilde{\mathbf{x}}^{n'}_{\lambda,k}$ was selected, $\prod_{u'=1}^{2} b^{n'}_{i'_\tau,i'_\nu,\lambda,u',k}$. Dividing the right-hand side by the constant $\prod_{n=1}^{N} p^n_0(\mathbf{y}^n_{\lambda,u,k}|\tilde{\mathbf{x}}^n_{\lambda,k})$:

$$\tilde{\beta}^{n'}_{\lambda,k} = \frac{p^{n'}_1(\mathbf{y}^{n'}_{\lambda,u,k}|\tilde{\mathbf{x}}^{n'}_{\lambda,k})}{p^n_0(\mathbf{y}^{n'}_{\lambda,u,k}|\tilde{\mathbf{x}}^{n'}_{\lambda,k}) \prod_{u'=1}^{2} b^{n'}_{i'_\tau,i'_\nu,\lambda,u',k}} p(\tilde{\mathbf{x}}^{n'}_{\lambda,k}|\tilde{\mathbf{x}}^{n'}_{\lambda,k-1}) \tag{5.18}$$

and then using (5.14), (5.15), (5.16) we have

$$\tilde{\beta}^{n'}_{\lambda,k} = \prod_{u=1}^{2} \breve{B}^{n'}_{\lambda,u,k} \frac{p^{n'}_1(\mathbf{y}^{n'}_{\lambda,3,k}|\tilde{\mathbf{x}}^{n'}_{\lambda,k})}{p^n_0(\mathbf{y}^{n'}_{\lambda,3,k}|\tilde{\mathbf{x}}^{n'}_{\lambda,k})} p(\tilde{\mathbf{x}}^{n'}_{\lambda,k}|\tilde{\mathbf{x}}^{n'}_{\lambda,k-1})$$

where the expressions for the likelihood are given in (5.3). We normalize $\{\tilde{\beta}^{n'}_{\lambda,k}\}^{N}_{n'=1}$ with the normalization factor

$$\tilde{B}_{\lambda,k} = \sum_{n'=1}^{N} \tilde{\beta}^{n'}_{\lambda,k} \tag{5.19}$$

And the normalized distribution is the following:

$$\tilde{b}^{n'}_{\lambda,k} = \frac{\tilde{\beta}^{n'}_{\lambda,k}}{\tilde{B}_{\lambda,k}}. \tag{5.20}$$

Next, we perform partition resampling, where we sample a partition index $j_{n'} \sim \{\tilde{b}^{n'}_{\lambda,k}\}^{N}_{n'=1}$ from the distribution of $\tilde{b}^{n'}_{\lambda,k}$ with replacement. The resulting selected partition has value $\mathbf{x}^{n}_{\lambda,k} = \tilde{\mathbf{x}}^{j_{n'}}_{\lambda,k}$ and selection probability $b^{n}_{\lambda,k} = \tilde{b}^{j_{n'}}_{\lambda,k}$. This concludes the independent partition likelihood-based proposal. We will proceed to combine these sampled partitions into particles that represent hypotheses on the entire multitarget state vector.

5.2.2 PARTICLE WEIGHTING

After partition resampling, we assemble particles from the sampled partitions as $\mathbf{X}^{n}_{k} = [\mathbf{x}^{n}_{1,k}{}^{T} \ldots \mathbf{x}^{n}_{L,k}{}^{T}]^{T}$. In particle weighting, we weight these particles with weights that incorporate prior and measurement information. We proceed to derive the exact expression of the weight equation.

We define a measurement matrix $\mathbf{Y}_k$ that is composed of measurements from $\mathbf{X}^{n}_{k}$ and contains measurements from each of the sensors u. Therefore,

$$\mathbf{Y}_k = \mathbf{y}^{n}_{u,k},\, n = 1, \ldots, N,\, u = 1, \ldots, U$$

where

$$\mathbf{y}^{n}_{u,k} = \left|\frac{1}{M}\sum_{m=0}^{M_d-1} d_{u,k}(m) \sum_{\lambda=1}^{\Lambda} s^{*}(m - \tau^{n}_{\lambda,u,k}) e^{-\frac{j2\pi m \nu^{n}_{\lambda,u,k}}{M}}\right|^{2} \Rightarrow$$

$$\begin{aligned}\mathbf{y}^{n}_{u,k} = \Bigg| &\sum_{l=1}^{L}\sum_{\lambda=1}^{\Lambda} A_{l,k} E_s \mathrm{AF}_s(\tau^{n}_{\lambda,u,k} - \tau_{l,u,k}, \nu_{l,u,k} - \nu^{n}_{\lambda,u,k}) \\ &+ \frac{1}{M}\sum_{m=0}^{M_d-1} v_{u,k}(m) \sum_{\lambda=1}^{\Lambda} s^{*}(m - \tau^{n}_{\lambda,u,k}) e^{-\frac{j2\pi m \nu^{n}_{\lambda,u,k}}{M}}\Bigg|^{2}.\end{aligned} \tag{5.21}$$

The likelihood function for each proposed particle $n' = 1, \ldots, N$ is given by

$$p^{n'}(\mathbf{y}^n_{u,k}|\mathbf{X}^1_k, \ldots, \mathbf{X}^{n'}_k, \ldots, \mathbf{X}^N_k), n = 1, \ldots, N, u = 1, \ldots, U \,.$$

Here, the hypothesis of particle n' is that the state equals $\mathbf{X}^{n'}_k$, while $\mathbf{y}^n_{u,k}, n = 1, \ldots, N, u = 1, \ldots, U$ are the measurements obtained from matched filters at the delay-Doppler location defined by the particle proposed target state vectors $\mathbf{X}^n_k, n = 1, \ldots, N$. This likelihood is, again, a multivariate exponential distribution [45] similarly to the single partition likelihood in (5.3). Using similar arguments as in the single partition case, we approximate the likelihood for each particle to

$$\prod_{u=1}^{U} p_1^{n'}(\mathbf{y}^{n'}_{u,k}|\mathbf{X}^{n'}_k) \prod_{n=1, n\neq n'}^{N} p_0^n(\mathbf{y}^n_{u,k}|\mathbf{X}^n_k)\,,$$

where $p_1^n(\mathbf{y}^n_{u,k}|\mathbf{X}^n_k)$ is the likelihood given that the target state equals $\mathbf{X}^n_k$ and $p_0^n(\mathbf{y}^n_k|\mathbf{X}^n_k)$ is the likelihood when no targets exist having state $\mathbf{X}^n_k$.

The weight of particle n' [4] using the assumptions above is given by

$$w_k^{n'} = \Gamma^{n'}_{k-1} \frac{\prod_{u=1}^{U} p_1^{n'}(\mathbf{y}^{n'}_{u,k}|\mathbf{X}^{n'}_k) \prod_{n=1,n\neq n'}^{N} p_0^n(\mathbf{y}^n_{u,k}|\mathbf{X}^n_k)}{\prod_{\lambda=1}^{\Lambda} b^{n'}_{\lambda,k}} p(\mathbf{X}^{n'}_k|\mathbf{X}^{n'}_{k-1})\,.$$

Dividing by the constant $\prod_{u=1}^{U}\prod_{n=1}^{N} p_0^n(\mathbf{y}^n_{u,k}|\mathbf{X}^n_k)$, we have

$$w_k^{n'} = \Gamma^{n'}_{k-1} \frac{\prod_{u=1}^{U} p_1^{n'}(\mathbf{y}^{n'}_{u,k}|\mathbf{X}^{n'}_k) p(\mathbf{X}^{n'}_k|\mathbf{X}^{n'}_{k-1})}{\prod_{u=1}^{U} p_0^{n'}(\mathbf{y}^{n'}_{u,k}|\mathbf{X}^{n'}_k) \prod_{\lambda=1}^{\Lambda} b^{n'}_{\lambda,k}}$$

where the expressions for the likelihood are given in (5.2). We normalize the above with $W_k = \sum_{n'=1}^{N} w_k^{n'}$ to obtain the normalized weighs

$$\Gamma^{n'}_k = \frac{\Gamma^{n'}_{k-1}}{W_k} \frac{\prod_{u=1}^{U} p_1^{n'}(\mathbf{y}^{n'}_{u,k}|\mathbf{X}^{n'}_k) p(\mathbf{X}^{n'}_k|\mathbf{X}^{n'}_{k-1})}{\prod_{u=1}^{U} p_0^{n'}(\mathbf{y}^{n'}_{u,k}|\mathbf{X}^{n'}_k) \prod_{\lambda=1}^{\Lambda} b^{n'}_{\lambda,k}} \tag{5.22}$$

and the estimate of the state as $\hat{\mathbf{X}}_k = \sum_{n'=1}^{N} \Gamma^{n'}_k \mathbf{X}^{n'}_k$. The algorithm is outlined in Table 5.1.

5.3 SCHEME FOR ADAPTIVE WAVEFORM SELECTION USING IPLPF

In order to improve tracking performance, we select at each time step k the set of parameters $\Theta_k = (Q, M, \kappa)$ of the MCPC CAZAC waveform to be transmitted so that a cost function based on the predicted mean squared tracking error is minimized. As shown below the predicted error will, apart from the adaptively selected Θ_k, depend on the random target strength vector $\mathbf{A}_k = [A_{l,k}]_{l=1}^{L}$, the random noise matrix $\mathbf{v}_k = [[v_{u,k}(m)]_{m=0}^{M-1}]_{u=1}^{U}$, and the unknown true target state $\mathbf{X}_k$. These are the random elements of the received waveform. Moreover, the tracking algorithm itself will introduce

Table 5.1: Independent Partition Likelihood Particle Filter

Likelihood Partition Sampling: Stage 1:

For each partition $\lambda = 1, \ldots, \Lambda$

For each particle $n = 1, \ldots, N$

Let $\check{\mathbf{x}}^n_{\lambda,k} = \mathbf{F}\mathbf{x}^n_{\lambda,k-1}$

For each sensor $u = 1, 2$ form

$\{\{\tau^n_{i_\tau,\lambda,u,k}, \nu^n_{i_\nu,\lambda,u,k}\}^{I^n_\tau}_{i_\tau=0}\}^{I^n_\nu}_{i_\nu=0}$

using (4.2), (5.5), (4.4), (4.5), (4.6), (5.9)

For $i_\tau = 0, \ldots, I^n_\tau$

For $i_\nu = 0, \ldots, I^n_\nu$

Evaluate $\mathbf{y}^n_{i_\tau,i_\nu,\lambda,u}$ using (5.12)

Evaluate $\check{b}^n_{i_\tau,i_\nu,\lambda,u}$ using (5.16)

Sample $j^n_{i_\tau,u}, j^n_{i_\nu,u} \sim \{\{\check{b}^n_{i_\tau,i_\nu,\lambda,u}\}^{I^n_\tau}_{i_\tau=0}\}^{I^n_\nu}_{i_\nu=0}$

Let $r^n_{i'_\tau,i'_\nu,\lambda,u,k} = \frac{c\tau^n_{j_{i_\tau},\lambda,u,k}T_b}{2}$,

$\dot{r}^n_{i'_\tau,i'_\nu,\lambda,u,k} = -\frac{c\nu^n_{j_{i_\nu},\lambda,u,k}\Delta\nu}{2f_c}$,

$b^n_{i'_\tau,i'_\nu,\lambda,u} = \check{b}^n_{j_{i_\tau},j_{i_\nu},\lambda,u}$

Calculate $\tilde{\mathbf{x}}^n_{\lambda,k}$ from $\{\tilde{r}^n_{i'_\tau,i'_\nu,\lambda,u,k}, \tilde{\dot{r}}^n_{i'_\tau,i'_\nu,\lambda,u,k}\}^2_{u=1}$

Likelihood Partition Sampling: Stage 2:

Evaluate $\mathbf{y}^{n'}_{\lambda,3,k}$ using (5.17)

Evaluate $\tilde{b}^{n'}_{\lambda,k}$ using (5.20)

Sample $j_{n'} \sim \{\tilde{b}^{n'}_{\lambda,k}\}^N_{n'=1}$

$\mathbf{x}^n_{\lambda,k} = \tilde{\mathbf{x}}^{j_{n'}}_{\lambda,k}$

$b^n_{\lambda,k} = \tilde{b}^{j_{n'}}_{\lambda,k}$

Particle Weighting:

Assemble particles $\mathbf{X}^n_k = [\mathbf{x}^n_{1,k} \ldots \mathbf{x}^n_{L,k}]$

Evaluate $\mathbf{Y}_k = \{\{\mathbf{y}^n_{u,k}\}^N_{n=1}\}^U_{u=1}$ using (5.21)

For each particle $n' = 1, \ldots, N$

Evaluate particle weights

$$\Gamma^{n'}_k = \frac{\Gamma^{n'}_{k-1}}{W_k}\frac{\prod_{u=1}^U p_1^{n'}(\mathbf{y}^{n'}_{u,k}|\mathbf{X}^{n'}_k)p(\mathbf{X}^{n'}_k|\mathbf{X}^{n'}_{k-1})}{\prod_{u=1}^U p_0^{n'}(\mathbf{y}^{n'}_{u,k}|\mathbf{X}^{n'}_k)\prod_{\lambda=1}^{\Lambda} b^{n'}_{\lambda,k}}$$

Estimation $\hat{\mathbf{X}}_k = \sum_{n'=1}^N \Gamma^{n'}_k \mathbf{X}^{n'}_k$

Increment k by 1

Adaptive selection of Θ_k

randomness. Specifically, the predicted error will also depend on the estimate of the multitarget state $\hat{\mathbf{X}}_k = \sum_{n=1}^{N} \Gamma_k^n \mathbf{X}_k^n$. This estimate depends on the set of particles $\{\mathbf{X}_k^n\}_{n=1}^N$ that will be sampled by the particle filter. A different set of particles, however, may be generated if we were to rerun time step k of the algorithm while keeping the true target state $\mathbf{X}_k$, target strengths $\mathbf{A}_k$, and noise terms $\mathbf{v}_k$ fixed. In other words, $\hat{\mathbf{X}}_k$ does not deterministically depend on the random elements of the received waveform and is treated as a random vector in the predicted error representation.

In the following analysis, we first derive the exact expected root mean squared error that can be used as a cost function for finding the optimal $\Theta_k = (Q, M, \kappa)$ in the RMSE sense. Next, using some simplifications and observations, we will construct a simpler cost function to significantly reduce the computational expense of the adaptive waveform selection process. The expected root mean squared error is given by the following:

$$J(\Theta_k) = E_{\hat{\mathbf{X}}_k, \mathbf{X}_k, \mathbf{A}_k, \mathbf{v}_k | \hat{\mathbf{X}}_{k-1}, \Theta_k}((\mathbf{X}_k - \hat{\mathbf{X}}_k)^T \mathbf{C}(\mathbf{X}_k - \hat{\mathbf{X}}_k))$$

where the weighting matrix $\mathbf{C}$ makes the units of the cost function consistent by compensating for the differing units of the state vector. In the above notation, we have placed the random elements $\hat{\mathbf{X}}_k, \mathbf{X}_k, \mathbf{A}_k, \mathbf{v}_k$ as subscripts. Moreover, the multitarget state estimate $\hat{\mathbf{X}}_{k-1}$ at $k-1$, given from the tracking process, and the choice of Θ_k are considered known and also placed as subscripts. The probability distributions of $\mathbf{A}_k$ and $\mathbf{v}_k$ were described in Section 5.1.2.

Next, we identify the set of values that the multitarget state estimate $\hat{\mathbf{X}}_k$ can take in terms of the delay-Doppler locations associated with it. As described in Section 5.2, in order to propose particles, we have considered a discrete finite set of delay-Doppler locations for each partition and each particle. This set corresponds to Cartesian coordinate locations that are most likely to occur according to the kinematic prior and the set of particles $\mathbf{X}_{k-1}^n, n = 1, \ldots, N$ generated at the previous time step $k-1$. The set is given in (5.10) as $\{\{\tau_{i_\tau,\lambda,u,k}^n, \nu_{i_\nu,\lambda,u,k}^n\}_{i_\tau=0}^{I_\tau^n}\}_{i_\nu=0}^{I_\nu^n}$, for $\lambda = 1, \ldots, \Lambda$, $u = 1, \ldots, 2$, and $n = 1, \ldots, N$. We use index $\jmath$ to denote a member of the set $\mathcal{G}$, of cardinality $|\mathcal{G}|$, consisting of the N particles that could be sampled by the IPLPF proposal and subsequently weighted. Therefore, $\mathcal{G}$ is a large set including all combinations of possible delay-Doppler locations from two of the sensors for each target and particle. The process of forming partitions from sampled delay-Doppler locations is explained in Section 5.2.1.1 and illustrated in Figure 5.1. Subsequently, one possible outcome of the likelihood sampling process and particle weighting is N weight-particle pairs $\{\Gamma_{\jmath,k}^n, \mathbf{X}_{\jmath,k}^n\}$, $n = 1, \ldots, N$ corresponding to delay-Doppler locations $\{\tau_{\jmath,\lambda,u,k}^n, \nu_{\jmath,\lambda,u,k}^n\}$, $\lambda = 1, \ldots, \Lambda$, $u = 1, \ldots, U$, $n = 1, \ldots, N$.

Similarly, based on the target motion model in this work, we can identify a discrete finite set of possible true target states $\mathbf{X}_k$. The set of likely true target locations in the AF plane is restricted due to prior information acquired during the tracking process and the discretization of the AF plane. Each possible true target state $\mathbf{X}_{\jmath',k}$ with index $\jmath'$ is a member of the set $\mathcal{G}'$, of cardinality $|\mathcal{G}'|$. $\mathbf{X}_{\jmath',k}$ is related to corresponding to delay-Doppler locations $\{\tau_{\jmath',l,u,k}, \nu_{\jmath',l,u,k}\}$, $l = 1, \ldots, L$, $u = 1, \ldots, U$.

From the above, we may rewrite the cost function as,

$$J(\Theta_k) = \int_{\mathbf{A}_k}\int_{\mathbf{v}_k} \sum_{j'=1}^{|\mathcal{G}'|}\sum_{j=1}^{|\mathcal{G}|}(\mathbf{X}_{j',k} - \sum_{n=1}^{N}\Gamma^n_{j,k}\mathbf{X}^n_{j,k})^T\mathbf{C}(\mathbf{X}_{j',k} - \sum_{n=1}^{N}\Gamma^n_{j,k}\mathbf{X}^n_{j,k})$$

$$p(\{\Gamma^n_{j,k}, \mathbf{X}^n_{j,k}\}_{n=1}^N, \mathbf{X}_{j',k}, \mathbf{A}_k, \mathbf{v}_k|\hat{\mathbf{X}}_{k-1}, \Theta_k)d\mathbf{A}_k d\mathbf{v}_k$$

The cost function is further expressed as

$$J(\Theta_k) = \int_{\mathbf{A}_k}\int_{\mathbf{v}_k} \sum_{j'=1}^{|\mathcal{G}'|}\sum_{j=1}^{|\mathcal{G}|}\sum_{n=1}^{N}\sum_{n'=1}^{N}(\mathbf{X}_{j',k} - \Gamma^n_{j,k}\mathbf{X}^n_{j,k})^T\mathbf{C}(\mathbf{X}_{j',k} - \Gamma^{n'}_{j,k}\mathbf{X}^{n'}_{j,k})$$

$$p(\{\Gamma^n_{j,k}, \mathbf{X}^n_{j,k}\}_{n=1}^N|\mathbf{X}_{j',k}, \mathbf{A}_k, \mathbf{v}_k, \hat{\mathbf{X}}_{k-1}, \Theta_k)$$

$$p(\mathbf{X}_{j',k}|\hat{\mathbf{X}}_{k-1})p(\mathbf{A}_k)p(\mathbf{v}_k)d\mathbf{A}_k d\mathbf{v}_k \; .$$

were the probability distributions $p(\mathbf{X}_{j',k}|\hat{\mathbf{X}}_{k-1})$, $p(\mathbf{A}_k)$, and $p(\mathbf{v}_k)$ are defined in the context of the motion and measurement models described in Sections 3.5.1, 5.1.2. Next, we examine the probability

$$p(\{\Gamma^n_{j,k}, \mathbf{X}^n_{j,k}\}_{n=1}^N|\mathbf{X}_{j',k}, \mathbf{A}_k, \mathbf{v}_k, \hat{\mathbf{X}}_{k-1}, \Theta_k)$$

that the jth set of particles $\{\mathbf{X}^n_{j,k}\}_{n=1}^N$ results from sampling by the IPLPF. To this end, we will follow the sampling process of the IPLPF and identify the selection probability for each partition of particles $\{\mathbf{X}^n_{j,k}\}_{n=1}^N$.

According to Section 5.2.1.1, we obtain values $\tilde{\mathbf{x}}^n_{\lambda,k}$ for each partition $\lambda = 1, \ldots, \Lambda$ and each particle $n = 1, \ldots, N$ by sampling delay-Doppler bins from sensors $u = 1, 2$ with probability $\prod_{u=1}^2 b^n_{i'_\tau,i'_\nu,\lambda,u,k}$ given by (5.16). The values $\tilde{\mathbf{x}}^n_{\lambda,k}$ allow us to evaluate the likelihoods for the U sensors in order to sample partitions. In Section 5.2.1.2, we obtain partitions $\mathbf{x}^n_{\lambda,k}$ with selection probability $b^n_{\lambda,k}$ given by (5.20). Next, in Section 5.2.2, we combine these sampled partitions into particles $\mathbf{X}^n_k$. From the sampling process of each particle $\mathbf{X}^n_k$, we conclude that the probability of each particle being selected is $\prod_{\lambda=1}^{\Lambda} b^n_{j,\lambda,k}\prod_{u=1}^2 b^n_{j,i'_\tau,i'_\nu,\lambda,u,k}$. Therefore, any set of particles $\{\mathbf{X}^n_{j,k}\}_{n=1}^N$ appears with probability $p(\{\Gamma^n_{j,k}, \mathbf{X}^n_{j,k}\}_{n=1}^N|\mathbf{X}_k, \mathbf{A}_k, \mathbf{v}_k, \{\mathbf{X}^n_{k-1}\}_{n=1}^N, \Theta_k) = \prod_{n=1}^N\prod_{\lambda=1}^{\Lambda} b^n_{j,\lambda,k}\prod_{u=1}^2 b^n_{j,i'_\tau,i'_\nu,\lambda,u,k}$. Further, using (5.3), (5.18), (5.19), (5.20)

$$\prod_{n=1}^{N}\prod_{\lambda=1}^{\Lambda} b^n_{j,\lambda,k}\prod_{u=1}^{2} b^n_{j,i'_\tau,i'_\nu,\lambda,u,k} =$$

$$\prod_{\lambda=1}^{\Lambda}\frac{\sigma^2_{\lambda,0}}{\tilde{B}_{\lambda,k}\sigma^2_{\lambda,1}} e^{\frac{\sigma^2_{\lambda,1}-\sigma^2_{\lambda,0}}{2\sigma^2_{\lambda,1}\sigma^2_{\lambda,0}}\sum_{n=1}^N\sum_{u=1}^U \mathbf{y}^n_{j,\lambda,u,k}} \prod_{n=1}^{N} p(\tilde{\mathbf{x}}^n_{j,\lambda,k}|\tilde{\mathbf{x}}^n_{\lambda,k-1}). \qquad (5.23)$$

where the matched filter statistic in (5.17), given specifically for the MCPC waveform with parameters Θ_k is

$$\mathbf{y}^n_{J,\lambda,u,k} = |\sum_{l=1}^{L} A_{l,k}\mathrm{AF}_{g_{\Theta_k}}(\tau^n_{J,\lambda,u,k} - \tau_{J',l,u,k}, \nu_{J',l,u,k} - \nu^n_{J,\lambda,u,k})$$

$$+ \sum_{m=0}^{M_d-1} v_{u,k}(m) g^*_{\Theta_k}(m - \tau^n_{J,\lambda,u,k}) e^{-\frac{j2\pi m \nu^n_{J,\lambda,u,k}}{M}}|^2. \tag{5.24}$$

Here, we note that since $\sigma^2_{\lambda,0} < \sigma^2_{\lambda,1}$, the selection probability is proportional to $\mathbf{y}^n_{n_p,\lambda,u,k}$.

The particle weights $\Gamma^n_{J,k}$ that are assigned to particles following the above sampling process also dependent on the AF, true target location, target strength, and noise terms. The particle weights are given in Section 5.2.2. Further using (5.2) and (5.22):

$$\Gamma^n_{J,k} = \frac{\Gamma^n_{k-1}}{W_k} \frac{p(\mathbf{X}^n_k|\mathbf{X}^n_{k-1})}{\prod_{\lambda=1}^{\Lambda} b^n_{\lambda,k}} \frac{\sigma^2_0}{\sigma^2_1} e^{\frac{\sigma^2_1 - \sigma^2_0}{2\sigma^2_1\sigma^2_0} \sum_{u=1}^{U} \mathbf{y}^n_{J,u,k}} \tag{5.25}$$

where using (5.21) the matched filter statistic for the MCPC waveform with parameters Θ_k is

$$\mathbf{y}^n_{J,u,k} = |\sum_{l=1}^{L}\sum_{\lambda=1}^{\Lambda} A_{l,k} E_s \mathrm{AF}_{g_{\Theta_k}}(\tau^n_{J,\lambda,u,k} - \tau_{J',l,u,k}, \nu_{J',l,u,k} - \nu^n_{J,\lambda,u,k})$$

$$+ \sum_{m=0}^{M_d-1} v_{u,k}(m) \sum_{\lambda=1}^{\Lambda} g^*_{\Theta_k}(m - \tau^n_{J,\lambda,u,k}) e^{-\frac{j2\pi m \nu^n_{J,\lambda,u,k}}{M}}|^2. \tag{5.26}$$

From the above, we observe that the cost function changes monotonically with a change in $\mathbf{A}_k$ or $\mathbf{v}_k$ for any value of the AF, and thus Θ_k. Therefore, the elements of the vectors $\mathbf{A}_k$ and $\mathbf{v}_k$ can be set to constants. If target strengths are known and are significantly different the best choice to replace $A_{l,k}$ with is the target strength $\sigma^2_{A,l}$, $l = 1, \ldots, L$. Since the noise level is assumed to be the same for all sensors and target returns, it can be set to zero for the purposes of minimizing the cost function. Therefore, the cost function can be approximated by

$$\bar{J}(\Theta_k) = \sum_{J'=1}^{|\mathcal{G}'|}\sum_{J=1}^{|\mathcal{G}|}\sum_{n=1}^{N}\sum_{n'=1}^{N} (\mathbf{X}_{J',k} - \Gamma^n_{J,k}\mathbf{X}^n_{J,k})^T \mathbf{C}(\mathbf{X}_{J',k} - \Gamma^{n'}_{J,k}\mathbf{X}^{n'}_{J,k})$$

$$p(\{\Gamma^n_{J,k}, \mathbf{X}^n_{J,k}\}_{n=1}^{N}|\mathbf{X}_{J',k}, \mathbf{A}_k, \hat{\mathbf{X}}_{k-1}, \Theta_k)$$

$$p(\mathbf{X}_{J',k}|\hat{\mathbf{X}}_{k-1}) .$$

The above expression can be further simplified if we assume a very large number of particles, which results to a more restricted set of *likely* weight-particle pairs to be produced with equal probability. This will eliminate the randomness introduced by the particle filter sampling process. Instead, we will content with some observations in order to construct a simpler cost function to significantly reduce computational complexity.

The first observation is that the error increases with an increase in AF sidelobes. This is due to the increase in $\Gamma^n_{J,k}$ and $p(\{\Gamma^n_{J,k}, \mathbf{X}^n_{J,k}\}_{n=1}^N | \{\sigma^2_{A,l}\}_{l=1}^L, \{\mathbf{X}^n_{k-1}\}_{n=1}^N, \Theta_k)$, which have a direct relationship with the AF as shown in (5.23), (5.24), (5.25), and (5.26) above. What enables the reduction of the expected error using adaptive waveform selection is that, for each time step, only a small set of delay-Doppler locations is interrogated by the particle filter. Therefore, the AF sidelobes at certain delay-Doppler locations that participate in the sampling process are minimized by the appropriate choice of the waveform parameters Θ_k. This results to an increase in sidelobes at other locations, preserving the uncertainty principle while not affecting the sampling process.

Overall, the delay-Doppler locations involved in the sampling and weighting process at each time step k are given by the difference of the proposed locations to the unknown true locations. This is due to the matched filter statistics construction in (5.24) and (5.26). The AF locations of interest are, therefore, given:

$$\Upsilon_k = \cup_{u=1}^U \cup_{n=1}^N \cup_{\lambda=1}^\Lambda \cup_{l=1}^L$$

$$\{\{\{(\tau^n_{\lambda,u,k,min} + i_\tau) - (\tau^n_{l,u,k,min} + i'_\tau)\}_{i_\tau=0}^{I^n_\tau}\}_{i'_\tau=0}^{I^n_\tau},$$

$$\{\{(\nu^n_{l,u,k,min} + i'_\nu) - (\nu^n_{\lambda,u,k,min} + i_\nu)\}_{i_\nu=0}^{I^n_\nu}\}_{i'_\nu=0}^{I^n_\nu}\} .$$

where $\tau^n_{\lambda,u,k,min} = \tau^n_{l,u,k,min}$ and $\nu^n_{l,u,k,min} = \nu^n_{\lambda,u,k,min}$ for $l = \lambda$.

Moreover, due to the term $(\mathbf{X}_{J',k} - \Gamma^n_{J,k}\mathbf{X}^n_{J,k})^T \mathbf{C}(\mathbf{X}_{J',k} - \Gamma^{n'}_{J,k}\mathbf{X}^{n'}_{J,k})$ an increase in $|\mathbf{X}_{J',k} - \Gamma^n_{J,k}\mathbf{X}^n_{J,k}|$ results to an increase in the cost function. What causes a particle hypothesis not being equal to the true target state ($\mathbf{X}_{J',k} \neq \mathbf{X}^n_{J,k}$) is again the non-zero AF sidelobes that appear at previous time steps in the iterative process of the particle filter. The term $(\mathbf{X}_{J',k} - \Gamma^n_{J,k}\mathbf{X}^n_{J,k})^T \mathbf{C}(\mathbf{X}_{J',k} - \Gamma^{n'}_{J,k}\mathbf{X}^{n'}_{J,k})$ additionally shows that sidelobes that are more distant to the origin of the AF contribute more to the error than sidelobes close to the origin.

In order to construct a simplified cost function, we take into account the AF sidelobes that lie in the AF regions interrogated by the particle filter (ignoring their distance from the origin of the AF) and the target strengths. The simplified cost function to be minimized is chosen in this work to be

$$\tilde{J}(\Theta_k) = \sum_{u=1}^U \sum_{n=1}^N \sum_{\lambda=1}^\Lambda \sum_{l=1}^L \sigma^2_{A,l} |\mathrm{AF}_{g_{\Theta_k}}(\tau^n_{J,\lambda,u,k} - \tau_{J',l,u,k}, \nu_{J',l,u,k} - \nu^n_{J,\lambda,u,k})|^2 .$$

In one of the scenarios considered in this paper, targets have significantly different strengths $\sigma^2_{A,l}$ for $l = 1, \ldots, L$. In this case, returns from strong targets mask returns from a weak target. This is due to the sum of terms $\sum_{\lambda=1}^{\Lambda} \sum_{l=1}^{L} \sigma^2_{A,l} \mathrm{AF}_{g_{\Theta_k}}(\tau^n_{J,\lambda,u,k} - \tau_{J',l,u,k}, \nu_{J',l,u,k} - \nu^n_{J,\lambda,u,k})$ appearing in the matched filter statistics in (5.24), and (5.26). Therefore, a proposed partition λ used to estimate a weak target will be weighted using a matched filter statistic that includes interference terms from all other $l = 1, \ldots, L, l \neq \lambda$ targets. These terms are given by $\sum_{l=1, l\neq\lambda}^{L} \sigma^2_{A,l} \mathrm{AF}_{g_{\Theta_k}}(\tau^n_{J,\lambda,u,k} - \tau_{J',l,u,k}, \nu_{J',l,u,k} - \nu^n_{J,\lambda,u,k})$. This means that an adaptive selected waveform should have low AF sidelobes at locations $\tau^n_{J,\lambda,u,k} - \tau_{l,u,k}, \nu_{l,u,k} - \nu^n_{J,\lambda,u,k}$ for $l = 1, \ldots, L$, where $l \neq \lambda$ in order to unmask weak targets.

Minimizing the abovementioned crossterms can be achieved by placing entire areas in the delay-Doppler plane where targets may exist in valleys of the AF expression that have been well defined in Section 2.3. The analysis performed in that section, therefore, facilitates the choice of parameters Θ_k, which minimize the cost function $\tilde{J}(\Theta_k)$.

5.4 SIMULATION RESULTS AND DISCUSSION

We use two scenarios, one with one weak target and two strong targets and another with all targets having equal strength to demonstrate the performance in tracking multiple targets with a) a SCPC Björck CAZAC, b) a non-adapted MCPC Björck CAZAC, and c) adaptively configured MCPC Björck CAZACs. Three targets move in a two-dimensional plane. The motion is completed in 199 time steps. Three sensors located at $\chi_1 = -1000$ m, $\psi_1 = 500$ m, $\chi_2 = 2500$ m, $\psi_2 = 500$ m, and $\chi_3 = 500$ m, $\psi_3 = 0$ m collect range and range rate measurements. The trajectory of the target and the location of the sensors are shown in Figure 5.2. The target is assumed to move according to a nearly constant velocity model with covariance matrix Q =diag(225 64 225 64).

For the first scenario, the weak target $l = 2$ has a cross-sectional area such that it is observed with SNR that varies as 5, 10, 12, 15, 17, 20 dB. These SNR values correspond to

$$\sigma^2_{A,2} = [3.16, 10.00, 15.85, 31.63, 50.12, 100.00]\,,$$

while we let the strong targets to be characterized by $\sigma^2_{A,1} = \sigma^2_{A,3} = \sigma^2_{A,2} + 1600$. The noise variance is set to $N_0 = 1$ and the energy of the waveform used is $E_s = 1$. In the second scenario, all three targets are observed with SNR that varies as 5, 10, 12, 15, 17, 20 dB.

The SCPC Björck CAZAC waveform has length $M = 1,741$. The choice of parameters of the MCPC waveforms was limited to combinations of values $\{M, Q\} = \{7, 245\}, \{11, 154\}, \{13, 130\}$, and $\zeta = 0, 1$ in order to reduce computational expense in the adaptive selection process. The Fourier transform of the waveform was also used to introduce another degree of freedom by rotating the AF. All waveforms are sampled at 8 MHz and frequency modulated by $f_c = 40$ GHz. The speed of propagation of the waveforms is $c = 2.997925 \times 10^8$ m/s. For the simulations, we used $N = 300$ particles, initialized by drawing from a Gaussian distribution with mean the true initial target position and covariance Q_0=diag(1000 100 1000 100). The results were averaged over 300 Monte

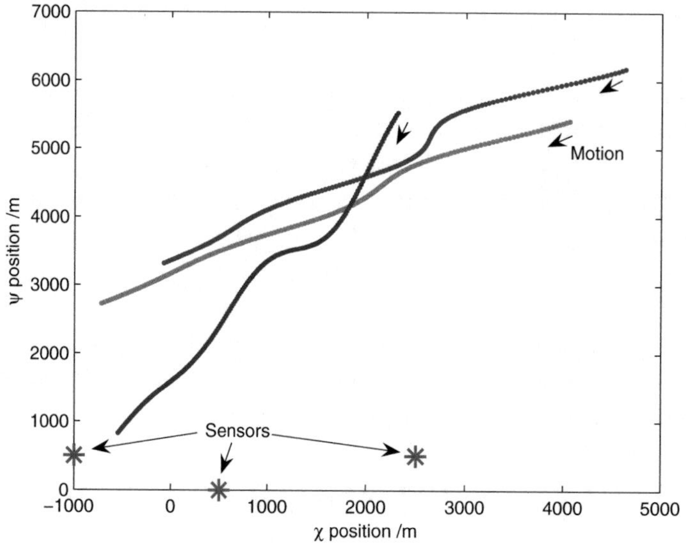

Figure 5.2: Target trajectory and sensor location.

Carlo runs. The parameters for the adaptively configured MCPC waveform are selected at each time step as described in Section 5.3, while for the fixed MCPC the parameters were selected randomly at the beginning of the scenario.

For the first scenario, the RMSE tracking performance is shown in Figure 5.3 for different values of SNR and for all waveforms. The percentage of lost tracks is shown for each waveform and SNR value in Figure 5.4. A lost track is declared if for 6 consecutive steps in the scenario, we have a tracking error of over 300m. We observe that the MCPC waveform with adaptive configuration (indicated as AMCPC in all figures) clearly outperforms the SCPC Björck CAZAC and fixed MCPC waveform when considering the number of lost tracks. In terms of the RMSE, the SCPC Björck CAZAC appears to have similar performance to the adaptive MCPC case since both have the same measurement resolution. Performing well in RMSE is, however, not useful if it is accompanied with a high number of lost tracks. The non-adaptive MCPC waveform case has the lowest performance rating due to its high sidelobes that are not avoided during measurement collection. In the case when there are no successful tracks, the RMSE value is shown as zero in Figures 5.3 and 5.5.

In the second scenario, from Figures 5.5 and 5.6, we observe that if the targets have equal strengths, then the adaptive MCPC and SCPC CAZAC perform similarly since their AF sidelobes do not mask weak targets. The large sidelobes of the non-adapted MCPC, however, still cause larger errors. Another observation is that in the first scenario, in the adaptive MCPC case the results are improved compared to the second scenario. This is because in the first scenario, two of the targets have higher SNR values than in the second scenario.

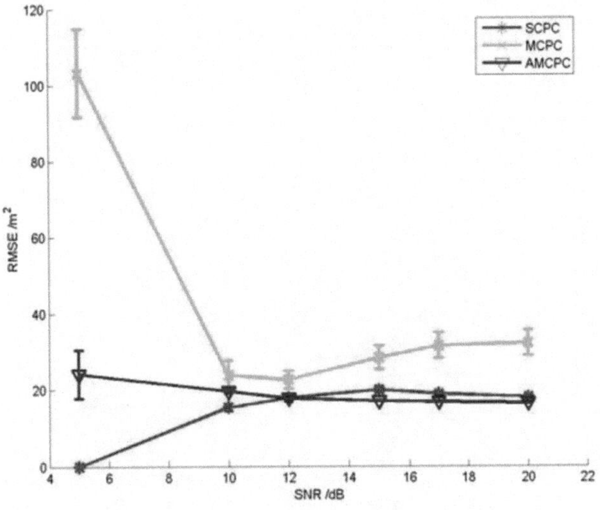

Figure 5.3: RMSE versus SNR for the three waveforms with 95% confidence intervals when tracking a weak target and two strong targets.

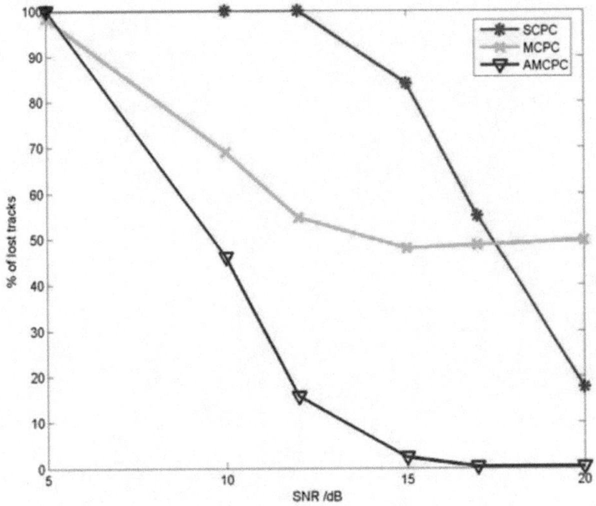

Figure 5.4: Percentage of lost tracks versus SNR for the three waveforms when tracking a weak target and two strong targets.

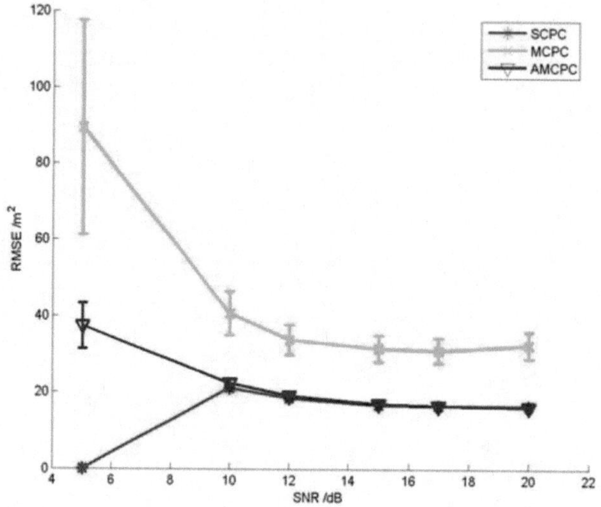

Figure 5.5: RMSE versus SNR for the three waveforms with 95% confidence intervals when tracking targets of equal strengths.

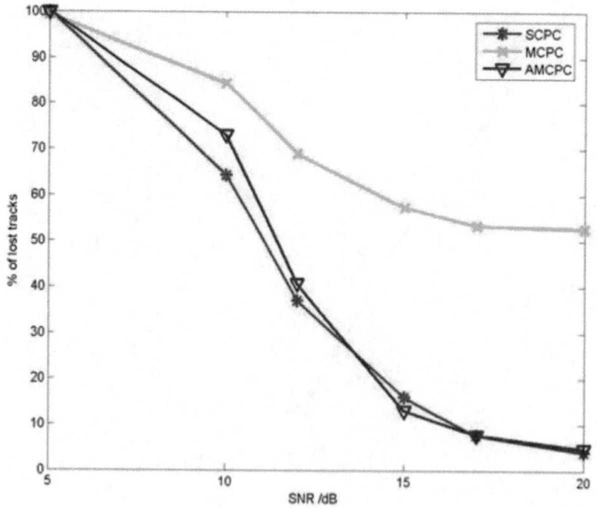

Figure 5.6: Percentage of lost tracks versus SNR for the three waveforms when tracking targets of equal strengths.

CHAPTER 6

Conclusions

In this work, we have described advances in target tracking using high resolution measurements and adaptive waveform selection. Specifically, we have outlined radar waveform processing methods, radar waveform designs, and an adaptive selection scheme. These methods utilize high resolution measurements effectively and reduce interference between target measurements in a multitarget environment in order to enhance tracking performance.

We have presented two particle filtering methods (the SIRPF and LPF) for tracking a single target using radar. Both particle filtering methods are able to deal with non-linear, non-Gaussian scenarios. Moreover, they concentrate their collection of measurements in areas where a target is believed to exist and avoid the exhaustive evaluation of matched filtering operations on a grid in the delay-Doppler space. As a consequence, the necessity to approximate probability of detection contours by tessellating shapes such as a parallelogram is avoided resulting to higher measurement accuracy. In addition, we have showed how the LPF algorithm is able to utilize the high-resolution measurements that are available when using the Björck CAZAC sequence.

Moreover, we have presented the IPLPF algorithm, a particle filtering method based on the independent partitions and the likelihood particle filter, for tracking a fixed and known number of targets. This particle filtering method can also deal with non-linear and non-Gaussian scenarios, while selecting measurements based on the belief on the target state. In addition, the IPLPF is able to utilize high resolution measurements resulting from the use of Björck CAZACs and MCPC Björck CAZAC sequences as radar waveforms and incorporate adaptive waveform selection. We showed how MCPC waveforms based on Björck CAZAC sequences exhibit AF surfaces with adaptively positioned sidelobes. We outlined a configuration strategy for selecting the parameter values of MCPC waveforms to position AF sidelobes such that weak targets are unmasked and the predicted mean squared tracking error is minimized. Finally, we have presented a simulation based study to assess the performance of the different radar system configurations. The simulation results demonstrate that when tracking targets with significantly different strengths the use of single Björck CAZAC waveforms or fixed-parameter MCPC waveforms results in deteriorated tracking performance. On the other hand, the use of adaptively configured MCPC waveforms enables the tracking of weak targets in the presence of strong targets and offers significant tracking performance gains.

APPENDIX A

Independence of Single Target Matched Filter Statistics

As explained in Chapter 4, the particle filtering formulation requires the evaluation of a multidimensional measurement likelihood for each of its particles. This is associated with a prohibitively high computational burden, especially when the dimensions of the state space and the number of particles increase. The remedy is to assume that the measurements are independent and that the multidimensional likelihood factorizes to each of its marginal components. The likelihood ratio to be evaluated will then become a simple expression.

In this appendix, we provide a general analytic expression of the multidimensional likelihood of the radar returns conditioned on particle-proposed target estimates. We examine the factors on which the measurement statistical dependence is attributed to, which gives an insight on when the measurement independence approximation is plausible. We illustrate the results with Monte Carlo (MC) generated plots of the covariance between measurements, which is related to measurement dependence, since the measurements are exponentially distributed. We plot the covariances for the cases of using a linear frequency modulated (LFM) [1] chirp and a Björck CAZAC [7] waveform. We show that for the Björck CAZAC, the measurement independence approximation is quite accurate due to its highly concentrated ambiguity function while for the LFM, this approximation does not hold as strongly.

We assume that at every time step k, a radar waveform $s(m)$, $m = 0, \ldots, M-1$, where M is the total number of samples of the waveform, is transmitted. From this point, we omit the subscripts k for ease of notation since we are investigating measurement independence irrespectively of the time step in the scenario. The return signal reflected from the target with a delay $\tau = 2r/c$ (where r is the range of the target with respect to the radar sensor and c is the velocity of propagation of the signals) and a Doppler $\nu = -\frac{2f_c\dot{r}_k}{c}$, with carrier frequency f_c is given:

$$d(m) = As(m-\tau)e^{j2\pi m\nu/M} + v(m),\ m = 0, \ldots, M-1 \tag{A.1}$$

where M is the total number of samples of the waveform. A is a sum of random complex returns from many different target scatterers, according to the Swerling I model [44]. Therefore, A is assumed to be zero-mean, complex Gaussian with known variance $2\sigma_A^2$ and $v(m)$ is zero-mean complex Gaussian noise with variance $2N_0$. We treat the return signal when no target is present as a special case of the above given signal, where $A = 0$. We further denote $A = A_I + jA_Q$, were the subscripts I and Q denote the in-phase and quadrature components, respectively. Similarly, $v(m) = v_I(m) + jv_Q(m)$.

The output of the matched filter at a proposed delay-Doppler location by a particle n is the following:

$$y^n = |\sum_{m=0}^{M-1} d(m)s^*(m-\tau^n)e^{-\frac{j2\pi m\nu^n}{M}}|^2 .$$

The likelihood function for each particle $n' = 1, \ldots, N$ is $p^{n'}(\mathbf{Y}|\mathbf{x}^1, \ldots, \mathbf{x}^{n'}, \ldots, \mathbf{x}^N)$, where $\mathbf{Y} = [y^1, \ldots, y^N]$ [5]. This vector of measurements has an N-dimensional exponential distribution [45]. Here, the hypothesis of particle n' is that the state equals $\mathbf{x}^{n'}$, while $\mathbf{Y}$ are the measurements obtained from matched filters at the delay-Doppler location defined by the particle proposed target state vectors $\{\mathbf{x}^n\}_{n=1}^N$. This likelihood appears in the weight equation of the particle filter described in Chapter 4.

Next, we examine the factors that affect the covariance between two measurements $y^{n'}, y^n$ with $n' \neq n$. Since we are dealing with a multivariate exponential distribution, zero covariance between measurements, implies independence.
Using (A.1) and the definition of the ambiguity function $\mathrm{AF}(\tau, \nu) = \sum_{m=0}^{M-1} s(m-\tau)e^{j2\pi m\nu/M}s^*(m)$, we obtain:

$$y^n = |A\mathrm{AF}(\tau^n - \tau, \nu - \nu^n) + \sum_{m=0}^{M-1} v(m)s^*(m-\tau^n)e^{-j2\pi\frac{m\nu^n}{M}}| .$$

Writing the above expression in terms of its in-phase and quadrature components, we obtain: $y^n = |\gamma_I^n + j\gamma_Q^n|$ where the in-phase and quadrature components are denoted, respectively, as

$$\gamma_I^n = \{A\mathrm{AF}(\tau^n - \tau, \nu - \nu^n)\}_I + \sum_{m=0}^{M-1} \{v(m)s^*(m-\tau^n)e^{-j2\pi\frac{m\nu^n}{M}}\}_I \quad \text{(A.2)}$$

$$\gamma_Q^n = \{A\mathrm{AF}(\tau^n - \tau, \nu - \nu^n)\}_Q + \sum_{m=0}^{M-1} \{v(m)s^*(m-\tau^n)e^{-j2\pi\frac{m\nu^n}{M}}\}_Q . \quad \text{(A.3)}$$

Here, γ_I^n and γ_Q^n are independent as $A_I, A_Q, v_I(m), v_Q(m)$ are independent zero-mean Gaussian random variables. Therefore, each y^n is exponentially distributed. We define the vectors $\Gamma_I = [\gamma_I^1 \gamma_I^2 \cdots \gamma_I^N]^T$ and $\Gamma_Q = \left[\gamma_Q^1 \gamma_Q^2 \cdots \gamma_Q^N\right]^T$. As in [45], we define matrices $\mathbf{K}_{I,I} = \mathbf{E}[\Gamma_I \Gamma_I^T]$, $\mathbf{K}_{Q,Q} = \mathbf{E}[\Gamma_Q \Gamma_Q^T]$, $\mathbf{K}_{I,Q} = \mathbf{E}[\Gamma_I \Gamma_Q^T]$, where $E[\cdot]$ denotes expectation. We also define $\mathbf{Y} = [y^1, \ldots, y^N]$; this vector of measurements has an N-dimensional exponential distribution [45] defined as the following:

$$p^{n'}(\mathbf{Y}|\mathbf{x}_k^1, \dots, \mathbf{x}_k^{n'}, \dots, \mathbf{x}_k^N) = \frac{1}{(4\pi)^N |\mathbf{K}|^{1/2}} \int_{-\pi}^{\pi} \dots \int_{-\pi}^{\pi} \exp(-\tfrac{1}{2} h(\mathbf{Y}, \Phi)) d\phi^1 \dots d\phi^N \tag{A.4}$$

where

$$\mathbf{K} = \begin{bmatrix} \mathbf{K}_{I,I} & \mathbf{K}_{I,Q} \\ \mathbf{K}_{I,Q}^T & \mathbf{K}_{Q,Q} \end{bmatrix}, \mathbf{K}^{-1} = \begin{bmatrix} \mathbf{B} & \mathbf{F} \\ \mathbf{F} & \mathbf{D} \end{bmatrix}$$

and

$$h(\mathbf{Y}, \Phi) = \sum_{n=1}^{N} (\mathbf{B}_{n,n} cos^2(\phi^n) + \mathbf{D}_{n,n} sin^2(\phi^n) + 2\mathbf{F}_{n,n} cos(\phi^n) sin(\phi^n)) y^n + \sum_{\substack{n,n'=1 \\ n \neq n'}}^{N} (\mathbf{B}_{n,n'} cos(\phi^n) cos(\phi^{n'}) + \mathbf{D}_{n,n'} sin(\phi^n) sin(\phi^{n'}) + 2\mathbf{F}_{n,n'} cos(\phi^n) sin(\phi^{n'}))(y^n y^{n'})^{(\frac{1}{2})}.$$

Also, $\Phi = [\phi^1 \phi^2 \dots \phi^N]^T$ with $\phi^n = \tan^{-1}(\frac{\gamma_Q^n}{\gamma_I^n})$ and

$$\mathbf{B} = (\mathbf{K}_{I,I} + \mathbf{K}_{I,Q} \mathbf{K}_{Q,Q}^{-1} \mathbf{K}_{I,Q}^T)^{-1}, \tag{A.5}$$

$$\mathbf{D} = (\mathbf{K}_{Q,Q} + \mathbf{K}_{I,Q}^T \mathbf{K}_{I,I}^{-1} \mathbf{K}_{I,Q}^T)^{-1}, \tag{A.6}$$

$$\mathbf{F} = -(\mathbf{K}_{I,I} - \mathbf{K}_{I,Q} \mathbf{K}_{Q,Q}^{-1} \mathbf{K}_{Q,I}^T)^{-1} \mathbf{K}_{I,Q} \mathbf{K}_{Q,Q}^{-1}. \tag{A.7}$$

Moreover, the covariance of y^n and $y^{n'}$ is given by

$$Cov(y^n y^{n'}) = 2((\mathbf{K}_{I,I})_{n,n'}^2 + (\mathbf{K}_{Q,Q})_{n,n'}^2 + (\mathbf{K}_{I,Q})_{n,n'}^2 + (\mathbf{K}_{Q,I})_{n,n'}^2). \tag{A.8}$$

We begin by analyzing $(\mathbf{K}_{I,I})_{n,n'}$ and $(\mathbf{K}_{Q,Q})_{n,n'}$:

$$(\mathbf{K}_{I,I})_{n,n'} = E[\{AAF(\tau^n - \tau, \nu - \nu^n)\}_I \{AAF(\tau^{n'} - \tau, \nu - \nu^{n'})\}_I + E[\sum_{m=0}^{M-1} \{v(m) s^*(m - \tau^n) e^{-j2\pi \frac{m\nu^n}{M}}\}_I \{v(m) s^*(m - \tau^{n'}) e^{-j2\pi \frac{m\nu^{n'}}{M}}\}_I].$$

Above, we used the independence of A_I and $v_I(m)$ and also the independence of $v_I(m)$, $v_I(m')$ for $m \neq m'$. Next, we simplify the first term of the summation of $(\mathbf{K}_{I,I})_{n,n'}$, using $A = A_I + jA_Q$, $\text{AF}(\tau^n - \tau, \nu - \nu^n) = \text{AF}_I(\tau^n - \tau, \nu - \nu^n) + j\text{AF}_Q(\tau^n - \tau, \nu - \nu^n)$, the independence of A_I and A_Q, and the fact that $\sigma_A^2 = E[A_I^2] = E[A_Q^2]$. Moreover, we simplify the second term using the relationships $v(m) = v_I(m) + jv_Q(m)$, the independence of $v_I(m)$ and $v_Q(m)$, and the relationships

$$s^*(m-\tau^n)e^{-j2\pi\frac{m\nu^n}{M}}s^*(m-\tau^{n'})e^{-j2\pi\frac{m\nu^{n'}}{M}} = \{s^*(m-\tau^n)e^{-j2\pi\frac{m\nu^n}{M}}s^*(m-\tau^{n'})e^{-j2\pi\frac{m\nu^{n'}}{M}}\}_I +$$

$$j\{s^*(m-\tau^n)e^{-j2\pi\frac{m\nu^n}{M}}s^*(m-\tau^{n'})e^{-j2\pi\frac{m\nu^{n'}}{M}}\}_Q\,,$$

and $N_0 = E[v_I^2(m)] = E[v_Q^2(m)]$. The result is the following:

$$\begin{aligned}(\mathbf{K}_{I,I})_{n,n'} =& \sigma_A^2 \text{AF}(\tau^n - \tau, \nu - \nu^n)\text{AF}^*(\tau^{n'} - \tau, \nu - \nu^{n'}) + \\ & N_0 \text{AF}_I(\tau^{n'} - \tau^n, \nu^n - \nu^{n'})\end{aligned} \tag{A.9}$$

Similarly, $(\mathbf{K}_{Q,Q})_{n,n'} = (\mathbf{K}_{I,I})_{n,n'}$. Moreover, we can easily obtain that $(\mathbf{K}_{I,Q})_{n,n'} = 0$ and $(\mathbf{K}_{Q,I})_{n,n'} = 0$, considering the independence of A_I and A_Q, A_I and $v_Q(m)$, A_Q and $v_I(m)$, and also $v_I(m)$ and $v_Q(m)$.

Therefore, using (A.8) and the above results, the covariance between two measurements induced from particles n and n' is $Cov(y^n y^{n'}) = 4(\mathbf{K}_{I,I})^2_{n,n'}$. Finally, we concentrate on the fact that the likelihood would factorize if $(\mathbf{K}_{I,I})_{n,n'} = 0, n \neq n'$. This would be true if the following happens:

$$\text{AF}(\tau^n - \tau, \nu - \nu^n)\text{AF}^*(\tau^{n'} - \tau, \nu - \nu^{n'}) = 0, \text{ and}$$

$$\text{AF}_I(\tau^{n'} - \tau^n, \nu^n - \nu^{n'}) = 0\,.$$

Therefore, the independence between two measurements depends on the distance between their matched filter locations with respect to the ambiguity function $\text{AF}_I(\tau^{n'} - \tau^n, \nu^n - \nu^{n'})$, and the covariance is amplified by N_0. In the presence of a target, the term $\text{AF}(\tau^n - \tau, \nu - \nu^n)\text{AF}^*(\tau^{n'} - \tau, \nu - \nu^{n'})$ becomes smaller as the particles move further from the target with respect to the spread of the ambiguity function, and the covariance is amplified by σ_A^2. Therefore, a strong target presence and a large noise covariance makes the measurements more correlated. What we can control is the waveform used in our radar application. A waveform with a narrow ambiguity function mainlobe and weak sidelobes would cause the covariance of two measurements to be low enough with a higher probability, causing them to be nearly independent.

Next, we demonstrate the factorization of the likelihood if $Cov(y^n y^{n'}) = 0$ for $n \neq n'$, implying that the measurements are independent. If we assume that $Cov(y^n y^{n'}) = 0$, therefore, $(\mathbf{K}_{I,I})_{n,n'} = 0, n \neq n'$, then from (B.5), (A.6), (A.7), we have $\mathbf{B} = \mathbf{D} = (\mathbf{K}_{I,I})^{-1}, \mathbf{B}_{n,n} = \frac{1}{(\mathbf{K}_{I,I})_{n,n}}, \mathbf{B}_{n,n'} = 0, n \neq n', \mathbf{F} = 0$. Therefore, (A.5) becomes

$$h(\mathbf{Y}, \Phi) = \sum_{n=1}^{N} \mathbf{B}_{n,n} y^n \tag{A.10}$$

with $\mathbf{B}_{n,n} = (\sigma_A^2 |\mathrm{AF}(\tau^n - \tau, \nu - \nu^n)|^2 + \frac{\sigma_0^2}{2})^{-1}$ since (A.9) with $n' = n$ is $(\mathbf{K}_{I,I})_{n,n} = \sigma_A^2 |\mathrm{AF}(\tau^n - \tau, \nu - \nu^n)|^2 + N_0 \mathrm{AF}(0, 0)$ and $\mathrm{AF}(0, 0) = 1$.

So far, we have assumed a target presence at a location proposed by any particle n'. Next, we also consider target absence on a location proposed by particle $n \neq n'$. If a target is present at n', but not n, then using $\sigma_1^2 = 2\sigma_A^2 |\mathrm{AF}(\tau^n - \tau, \nu - \nu^n)|^2 + \sigma_0^2$ [2] then $\mathbf{B}_{n',n'} = \frac{2}{\sigma_1^2}, \mathbf{B}_{n,n} = \frac{2}{\sigma_0^2}$. Using (A.10) and the expressions for $\mathbf{B}_{n',n'}$ and $\mathbf{B}_{n,n}$, (A.4) becomes the following:

$$\begin{aligned} & p^{n'}(\mathbf{Y}|\mathbf{x}^1, \ldots, \mathbf{x}^{n'}, \ldots, \mathbf{x}^N) = \\ & \frac{1}{(4\pi)^N |K|^{1/2}} \exp(-\frac{y^{n'}}{\sigma_1^2}) \prod_{\substack{n=1 \\ n \neq n'}}^{N} \exp(-\frac{y^n}{\sigma_0^2}). \end{aligned} \tag{A.11}$$

We emphasize that in the above expression, the assumption is that a single target is present at a location $\mathbf{x}^{n'}$ proposed by target n', and it is not present in any location $\mathbf{x}^n$ proposed by the rest of the particles $n = 1, \ldots, N, n \neq n'$. The likelihood ratio used in the particle filter weight equation is the ratio of the likelihood when a target is present at $\mathbf{x}^{n'}$ to the likelihood when a target is absent. This is given by the simple expression:

$$\frac{p^{n'}(\mathbf{Y}|\mathbf{x}^1, \ldots, \mathbf{x}^{n'}, \ldots, \mathbf{x}^N)}{p_0^{n'}(\mathbf{Y}|\mathbf{x}^1, \ldots, \mathbf{x}^{n'}, \ldots, \mathbf{x}^N)} = \frac{\exp(-\frac{y^{n'}}{\sigma_1^2})}{\exp(-\frac{y^{n'}}{\sigma_0^2})}.$$

We provide numerical values for the covariance of the measurements when using the LFM and Björck CAZAC, in the cases where a single target is present or absent. The plots in this appendix illustrate the covariances, averaged over 100 MC simulations, between two measurements taken in different locations of the time-frequency plane. A target (if it exists) is located at $(0, 0)$ of the delay-Doppler plane. We vary the location of a particle n' at different bins in the delay-Doppler plane, and for each location, we plot the value of the cross-covariance of the particle n' with a reference particle n at a fixed location.

In Figures A.1 and A.2, we fix the location of a particle n at $(0, 0)$ and place a particle n' at all other possible locations of the delay-Doppler plane. The target is present at $(0, 0)$. In Figure A.1, we use the LFM, while in Figure A.2, we use the Björck CAZAC.

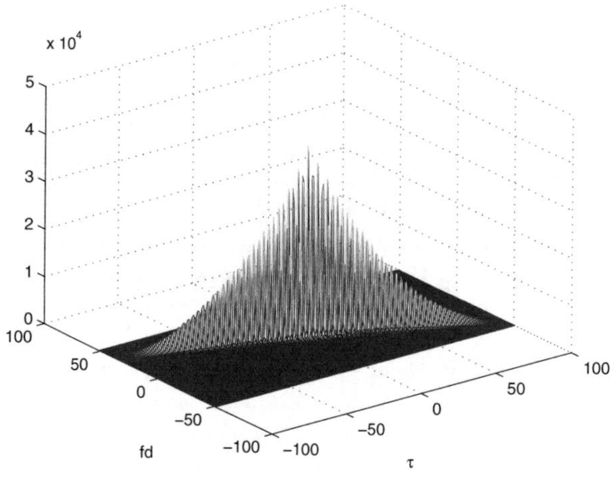

Figure A.1: Particle n at (0, 0) and a target is present at (0, 0). An LFM waveform is used.

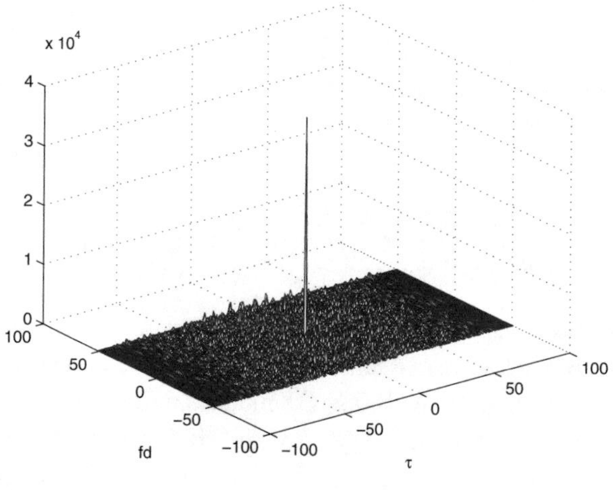

Figure A.2: Particle n at (0, 0) and a target is present at (0, 0). A CAZAC waveform is used.

In Figures A.3 and A.4 we place a particle n at a far end of the delay-Doppler plane (100, 50) and a particle n' at all other possible locations. The target is present at (0, 0). We use the LFM in Figure A.3 and the Björck CAZAC in Figure A.4.

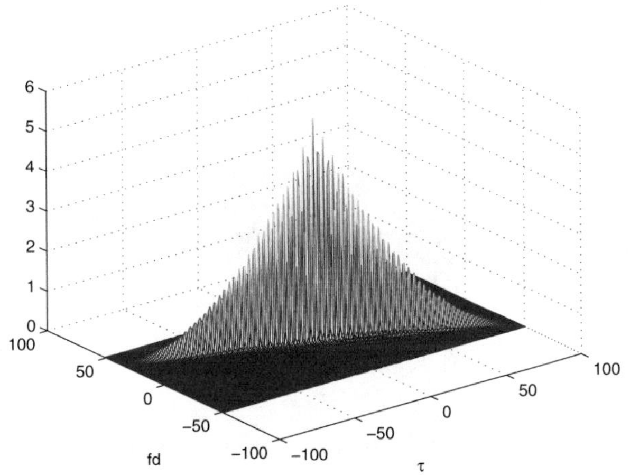

Figure A.3: Particle n at (100, 50) and a target is present at (0, 0). An LFM waveform is used.

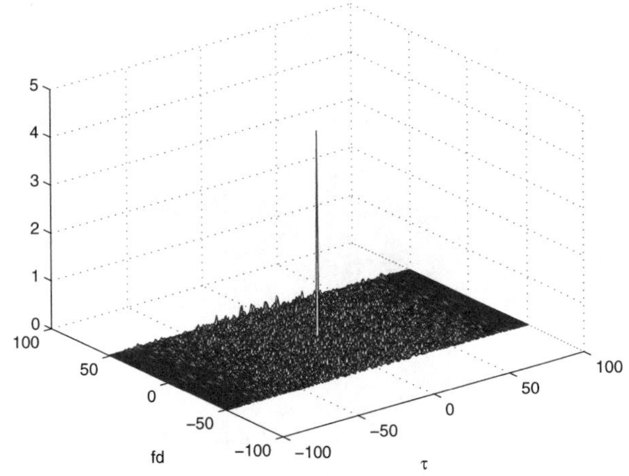

Figure A.4: Particle n at (100, 50) and a target is present at (0, 0). A CAZAC waveform is used.

In Figures A.5 and A.6, we place a particle n at $(0, 0)$ and a particle n' at all other possible locations. In this case, there is no target present. We use the LFM in Figure A.5 and the Björck CAZAC Figure A.6.

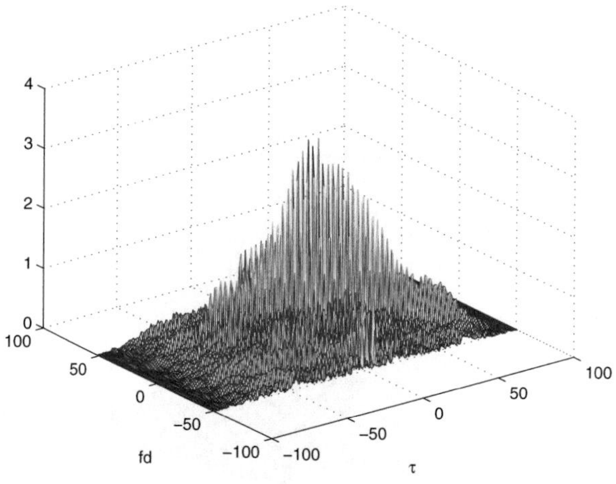

Figure A.5: Particle n at $(0, 0)$ and a target is not present. An LFM waveform is used.

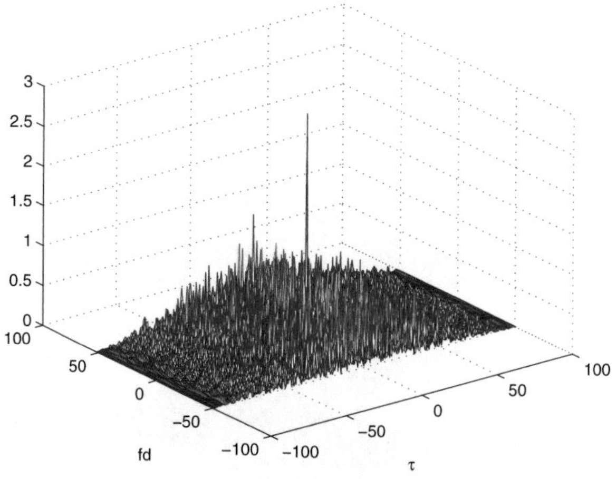

Figure A.6: Particle n at $(0, 0)$ and a target is not present. A CAZAC waveform is used.

In all plots the signal-to-noise ratio (SNR) is assumed to be 20 dB. The length of the waveforms used is 101 samples. It is evident in all the plots that using a Björck CAZAC makes the independence assumption more reasonable as the covariance of two measurements degrades rapidly when the measurements are taken further from each other and the target. We note that the ambiguity function of the LFM, and therefore, its covariance plot is more spread than that of the CAZAC for all possible chirp rate parameters of the LFM. The results clearly indicate that the thumbtack ambiguity function of the Björck CAZAC makes the measurement independence assumption plausible. For the LFM, however, the measurement dependence decreases slowly across the ridge of its ambiguity function. Therefore, using the Björck CAZAC is recommended for use with a particle filter based algorithm.

APPENDIX B

Independence of Multiple Target Matched Filter Statistics

In this appendix, we provide the analytic expression of the multidimensional likelihood of the radar returns conditioned on particle-proposed target estimates for the case when we have multiple targets. This multidimensional likelihood is used in Chapter 5 to evaluate the weights of particles in the IPLPF algorithm. Assuming statistical independence between measurements simplifies the computation of weights greatly reducing computational expense. We examine the factors on which the measurement statistical dependence is attributed to, which gives an insight on when the measurement independence approximation is plausible. We show that for the both, the single and MCPC Björck CAZAC waveform the measurement independence approximation is quite accurate due to their highly concentrated AF at delay-Doppler areas from which we choose to extract measurements.

We start by examining the factors that affect the covariance between two measurements $\mathbf{y}_{u,k}^{n'}$, $\mathbf{y}_{u,k}^{n}$ with $n' \neq n$. Since we are dealing with a multivariate exponential distribution, zero covariance between measurements, implies independence. Next, we exclude the use of subscripts denoting the sensor index and time-step, u and k respectively, for simplicity. The return waveform at the receiver is given:

$$d(m) = \sum_{l=1}^{L} A_l s(m - \tau_l) e^{\frac{j2\pi m \nu_l}{M}} + v(m)$$

where A_l is a sum of random complex returns from many different target scatterers on target l, according to the Swerling I model [44]. Therefore, A_l is assumed to be zero-mean, complex Gaussian with known variance $2\sigma_{A,l}^2$. The noise term $v(m)$ is assumed to be zero-mean complex Gaussian with variance $2N_0$.

The return signal is filtered to match a template signal representing returns from Λ targets, at different delay and Doppler locations $\tilde{\tau}_\lambda$, $\tilde{\nu}_\lambda$, $\lambda = 1, \ldots, \Lambda$, respectively. These delay and Doppler locations are derived from the belief in target state as represented by the particles of a particle filtering approach as described in Chapter 5. Given a set of delay and Doppler values $\{\tilde{\tau}_\lambda\}$ and $\{\tilde{\nu}_\lambda\}$, the matched filter output $\tilde{\mathbf{y}}$ is formed:

$$\tilde{\mathbf{y}} = \sum_{l=1}^{L} \sum_{\lambda=1}^{\Lambda} A_l E_s \mathrm{AF}_s(\tilde{\tau}_\lambda - \tau_l, \nu_l - \tilde{\nu}_\lambda) +$$

$$\sum_{m=0}^{M_d-1} v(m) \sum_{\lambda=1}^{\Lambda} s^*(m-\tilde{\tau}_\lambda) e^{-\frac{j2\pi m\tilde{\nu}_\lambda}{M}} .$$

The matched filter statistic that we will use for estimation is given by $\mathbf{y} = |\tilde{\mathbf{y}}|^2$.

$$\mathbf{y}^n = |\sum_{l=1}^{L}\sum_{\lambda=1}^{\Lambda} A_l E_s \mathrm{AF}(\tau_\lambda^n - \tau_l, \nu_l - \nu_\lambda^n) + \sum_{m=0}^{M_d-1} v(m) \sum_{\lambda=1}^{\Lambda} s^*(m-\tau_\lambda^n) e^{-\frac{j2\pi m\nu_\lambda^n}{M}}|^2. \tag{B.1}$$

Writing the above expression in terms of its in-phase and quadrature components, we obtain: $\mathbf{y}^n = |\gamma_I^n + j\gamma_Q^n|^2$ where the in-phase and quadrature components are denoted, respectively, as

$$\begin{aligned}\gamma_I^n = &\sum_{l=1}^{L}\sum_{\lambda=1}^{\Lambda} E_s\{A_l \mathrm{AF}(\tau_\lambda^n - \tau_l, \nu_l - \nu_\lambda^n)\}_I + \\ &+ \sum_{m=0}^{M_d-1} \{v(m) \sum_{\lambda=1}^{L'} s^*(m-\tau_\lambda^n) w^{-\frac{j2\pi m\nu_\lambda^n}{M}}\}_I\end{aligned} \tag{B.2}$$

$$\begin{aligned}\gamma_Q^n = &\sum_{l=1}^{L}\sum_{\lambda=1}^{\Lambda} E_s\{A_l \mathrm{AF}(\tau_\lambda^n - \tau_l, \nu_l - \nu_\lambda^n)\}_Q + \\ &+ \sum_{m=0}^{M_d-1} \{v(m) \sum_{\lambda=1}^{\Lambda} s^*(m-\tau_\lambda^n) e^{-\frac{j2\pi m\nu_\lambda^n}{M}}\}_Q.\end{aligned} \tag{B.3}$$

Here, γ_I^n and γ_Q^n are independent as $A_{l,I}, A_{l,Q}, v_I(m), v_Q(m)$ are independent zero-mean Gaussian random variables. Therefore, each $\mathbf{y}^n$ is exponentially distributed. We define the vectors $\Gamma_I = \left[\gamma_I^1 \gamma_I^2 \dots \gamma_I^N\right]^T$ and $\Gamma_Q = \left[\gamma_Q^1 \gamma_Q^2 \dots \gamma_Q^N\right]^T$. As in [45], we define matrices $\mathbf{K}_{I,I} = \mathbf{E}[\Gamma_I \Gamma_I^T]$, $\mathbf{K}_{Q,Q} = \mathbf{E}[\Gamma_Q \Gamma_Q^T]$, $\mathbf{K}_{I,Q} = \mathbf{E}[\Gamma_I \Gamma_Q^T]$, where $E[\cdot]$ denotes expectation. We also define $\mathbf{Y} = [\mathbf{y}^1, \dots, \mathbf{y}^N]$; this vector of measurements has an N-dimensional exponential distribution [45] defined as the following:

$$\begin{aligned}&p^{n'}(\mathbf{Y}|\mathbf{x}^1, \dots, \mathbf{x}^{n'}, \dots, \mathbf{x}^N) = \\ &\frac{1}{(4\pi)^N |\mathbf{K}|^{1/2}} \int_{-\pi}^{\pi} \dots \int_{-\pi}^{\pi} \exp(-\tfrac{1}{2} h(\mathbf{Y}, \Phi)) d\phi^1 \dots d\phi^N\end{aligned} \tag{B.4}$$

where

$$\mathbf{K} = \begin{bmatrix} \mathbf{K}_{I,I} & \mathbf{K}_{I,Q} \\ \mathbf{K}_{I,Q}^T & \mathbf{K}_{Q,Q} \end{bmatrix}, \mathbf{K}^{-1} = \begin{bmatrix} \mathbf{B} & \mathbf{F} \\ \mathbf{F} & \mathbf{D} \end{bmatrix}$$

and

$$h(\mathbf{Y},\Phi)=\sum_{n=1}^{N}(\mathbf{B}_{n,n}cos^2(\phi^n)+\mathbf{D}_{n,n}sin^2(\phi^n)+$$
$$2\mathbf{F}_{n,n}cos(\phi^n)sin(\phi^n))\mathbf{y}^n+\sum_{\substack{n,n'=1\\n\neq n'}}^{N}(\mathbf{B}_{n,n'}cos(\phi^n)cos(\phi^{n'})+$$
$$\mathbf{D}_{n,n'}sin(\phi^n)sin(\phi^{n'})+2\mathbf{F}_{n,n'}cos(\phi^n)sin(\phi^{n'}))(\mathbf{y}^n\mathbf{y}^{n'})^{(\frac{1}{2})}.$$

Also, $\Phi=[\phi^1\phi^2\ldots\phi^N]^T$ with $\phi^n=\tan^{-1}(\frac{\gamma_Q^n}{\gamma_I^n})$ and

$$\mathbf{B}=(\mathbf{K}_{I,I}+\mathbf{K}_{I,Q}\mathbf{K}_{Q,Q}^{-1}\mathbf{K}_{I,Q}^T)^{-1}, \quad \text{(B.5)}$$
$$\mathbf{D}=(\mathbf{K}_{Q,Q}+\mathbf{K}_{I,Q}^T\mathbf{K}_{I,I}^{-1}\mathbf{K}_{I,Q}^T)^{-1}, \quad \text{(B.6)}$$
$$\mathbf{F}=-(\mathbf{K}_{I,I}-\mathbf{K}_{I,Q}\mathbf{K}_{Q,Q}^{-1}\mathbf{K}_{Q,I}^T)^{-1}\mathbf{K}_{I,Q}\mathbf{K}_{Q,Q}^{-1}. \quad \text{(B.7)}$$

Moreover, the covariance of $\mathbf{y}^n$ and $\mathbf{y}^{n'}$ is given by

$$Cov(\mathbf{y}^n\mathbf{y}^{n'})=2((\mathbf{K}_{I,I})_{n,n'}^2+(\mathbf{K}_{Q,Q})_{n,n'}^2+(\mathbf{K}_{I,Q})_{n,n'}^2+(\mathbf{K}_{Q,I})_{n,n'}^2). \quad \text{(B.8)}$$

We begin by analyzing $(\mathbf{K}_{I,I})_{n,n'}$ and $(\mathbf{K}_{Q,Q})_{n,n'}$:

$$(\mathbf{K}_{I,I})_{n,n'}=E[E_s\sum_{l=1}^{L}\sum_{\lambda=1}^{\Lambda}\{A_l\text{AF}(\tau_\lambda^n-\tau_l,\nu_l-\nu_\lambda^n)\}_I$$
$$E_s\sum_{l=1}^{L}\sum_{\lambda'=1}^{\Lambda}\{A_l\text{AF}(\tau_{\lambda'}^{n'}-\tau_l,\nu_l-\nu_{\lambda'}^{n'})\}_I+$$
$$E[\sum_{m=0}^{M_d-1}\{v(m)\sum_{\lambda=1}^{\Lambda}s^*(m-\tau_\lambda^n)e^{-\frac{j2\pi m\nu_\lambda^n}{M}}\}_I$$
$$\sum_{m=0}^{M_d-1}\{v(m)\sum_{\lambda'=1}^{\Lambda}s^*(m-\tau_{\lambda'}^{n'})e^{-\frac{j2\pi m\nu_{\lambda'}^{n'}}{M}}\}_I].$$

Above, we used the independence of α_I and $v_I(m)$ and also the independence of $v_I(m)$, $v_I(m')$ for $m\neq m'$. Next, we simplify the above expression using $A_l=A_{l,I}+jA_{l,Q}$, $\text{AF}(\tau_\lambda^n-\tau_l,\nu_l-\nu_\lambda^n)=\text{AF}_I(\tau_\lambda^n-\tau_l,\nu_l-\nu_\lambda^n)+j\text{AF}_Q(\tau_\lambda^n-\tau_l,\nu_l-\nu_\lambda^n)$, the independence of $A_{l,I}$ and $A_{l,Q}$, and the fact that $\sigma_{A,l}^2=E[A_{l,I}^2]=E[A_{l,Q}^2]$. Moreover, we simplify the noise terms using the relationships $v(m)=v_I(m)+jv_Q(m)$, the independence of $v_I(m)$ and $v_Q(m)$, and the relationships $s^*(m-\tau_\lambda^n)e^{-\frac{j2\pi m\nu_\lambda^n}{M}} = \{s^*(m-\tau_\lambda^n)e^{-\frac{j2\pi m\nu_\lambda^n}{M}}\}_I + j\{s^*(m-\tau_\lambda^n)e^{-\frac{j2\pi m\nu_\lambda^n}{M}}\}_Q$, and $N_0 = E[v_I^2(m)] = E[v_Q^2(m)]$. The result is:

$$(\mathbf{K}_{I,I})_{n,n'} = E_s^2 \sum_{l=1}^{L} \sigma_{A,l}^2 \sum_{\lambda=1}^{\Lambda}\sum_{\lambda'=1}^{\Lambda} \mathrm{AF}(\tau_\lambda^n - \tau_l, \nu_l - \nu_\lambda^n)\mathrm{AF}^*(\tau_{\lambda'}^{n'} - \tau_l, \nu_l - \nu_{\lambda'}^{n'}) + N_0 E_s \sum_{\lambda=1}^{\Lambda}\sum_{\lambda'=1}^{\Lambda} \mathrm{AF}_I(\tau_\lambda^n - \tau_{\lambda'}^{n'}, \nu_{\lambda'}^{n'} - \nu_{l'}^n). \tag{B.9}$$

Similarly, we find that $(\mathbf{K}_{Q,Q})_{n,n'} = (\mathbf{K}_{I,I})_{n,n'}$. Moreover, considering the independence of α_I and α_Q, α_I and $v_Q(m)$, α_Q and $v_I(m)$, and also $v_I(m)$ and $v_Q(m)$, we can easily obtain that $(\mathbf{K}_{I,Q})_{n,n'} = 0$ and $(\mathbf{K}_{Q,I})_{n,n'} = 0$.

Therefore, using (A.8) and the above results, the covariance between two measurements induced from particles n and n' is $Cov(\mathbf{y}^n\mathbf{y}^{n'}) = 4(\mathbf{K}_{I,I})^2_{n,n'}$. Finally, we concentrate on the fact that the likelihood would factorize if $(\mathbf{K}_{I,I})_{n,n'} = 0, n \neq n'$. This would be true if the following happens:

$$\sum_{\lambda=1}^{\Lambda}\sum_{\lambda'=1}^{\Lambda} \mathrm{AF}(\tau_\lambda^n - \tau_l, \nu_l - \nu_\lambda^n)\mathrm{AF}^*(\tau_{\lambda'}^{n'} - \tau_l, \nu_l - \nu_{\lambda'}^{n'}) = 0, \text{ and}$$

$$\sum_{\lambda=1}^{\Lambda}\sum_{\lambda'=1}^{\Lambda} \mathrm{AF}_I(\tau_\lambda^n - \tau_{\lambda'}^{n'}, \nu_{\lambda'}^{n'} - \nu_\lambda^n) = 0\,.$$

Therefore, the independence between two measurements from particles n and n' depends on the distance between their matched filter locations with respect to the AF $\mathrm{AF}_I(\tau_\lambda^n - \tau_{\lambda'}^{n'}, \nu_{\lambda'}^{n'} - \nu_\lambda^n)$, and the covariance is amplified by N_0. In the presence of a target, the term $\mathrm{AF}(\tau_\lambda^n - \tau_l, \nu_l - \nu_\lambda^n)\mathrm{AF}^*(\tau_{\lambda'}^{n'} - \tau_l, \nu_l - \nu_{\lambda'}^{n'})$ becomes smaller as the particles move further from the target with respect to the spread of the AF, and the covariance is amplified by $\sigma_{A,l}^2$. Therefore, a strong target presence and a large noise covariance makes the measurements more correlated. What we can control is the waveform used in our radar application. The use of the Björck CAZAC with a narrow AF mainlobe and weak sidelobes causes the covariance of two measurements to be low enough with a higher probability, causing them to be nearly independent. The same assumption as for the single Björck CAZAC holds for adaptively configured MCPC Björck CAZAC waveforms. Although MCPC Björck CAZACs exhibit large sidelobes at certain locations of the AF, the adaptive configuration described in Chapter 5 ensures that the areas of the AF used in tracking have mostly zero sidelobes.

Next, we demonstrate the factorization of the likelihood if $Cov(\mathbf{y}^n\mathbf{y}^{n'}) = 0$ for $n \neq n'$, implying that the measurements are independent. If we assume that $Cov(\mathbf{y}^n\mathbf{y}^{n'}) = 0$, therefore, $(\mathbf{K}_{I,I})_{n,n'} = 0, n \neq n'$; then from (B.5), (A.6), (A.7), we have $\mathbf{B} = \mathbf{D} = (\mathbf{K}_{I,I})^{-1}, \mathbf{B}_{n,n} = \frac{1}{(\mathbf{K}_{I,I})_{n,n}}, \mathbf{B}_{n,n'} = 0, n \neq n', \mathbf{F} = 0$. Therefore, (A.5) becomes

$$h(\mathbf{Y}, \Phi) = \sum_{n=1}^{N} \mathbf{B}_{n,n}\mathbf{y}^n \tag{B.10}$$

with $\mathbf{B}_{n,n} = (E_s^2 \sum_{l=1}^{L} \sigma_{A,l}^2 \sum_{\lambda=1}^{\Lambda} \sum_{\lambda'=1}^{\Lambda} \text{AF}(\tau_\lambda^n - \tau_l, \nu_l - \nu_\lambda^n)\text{AF}^*(\tau_{\lambda'}^n - \tau_l, \nu_l - \nu_{\lambda'}^n) + N_0 E_s \sum_{\lambda=1}^{\Lambda} \sum_{\lambda'=1}^{\Lambda} \text{AF}_I(\tau_\lambda^n - \tau_{\lambda'}^n, \nu_{\lambda'}^n - \nu_\lambda^n))^{-1}$ from (A.9) with $n' = n$ and $\text{AF}(0, 0) = 1$.

So far, we have assumed a target presence at a location proposed by any particle n'. Next, we also consider target absence on a location proposed by particle $n \neq n'$. If a target is present at n', but not n, then using $\sigma_1^2 = 2E_s^2 \sum_{l=1}^{L} \sigma_{A,l}^2 \sum_{\lambda=1}^{\Lambda} \sum_{\lambda'=1}^{\Lambda} \text{AF}(\tau_\lambda^n - \tau_l, \nu_l - \nu_\lambda^n)\text{AF}^*(\tau_{\lambda'}^n - \tau_l, \nu_l - \nu_{\lambda'}^n) + 2N_0 E_s \sum_{\lambda=1}^{\Lambda} \sum_{\lambda'=1}^{\Lambda} \text{AF}_I(\tau_\lambda^n - \tau_{\lambda'}^n, \nu_{\lambda'}^n - \nu_\lambda^n)$ and $\sigma_0^2 = 2N_0 E_s \sum_{\lambda=1}^{\Lambda} \sum_{\lambda'=1}^{\Lambda} \text{AF}(\tilde{\tau}_\lambda - \tilde{\tau}_{\lambda'}, \tilde{\nu}_{\lambda'} - \tilde{\nu}_{\lambda'})$ from Chapter 5, then $\mathbf{B}_{n',n'} = \frac{2}{\sigma_1^2}, \mathbf{B}_{n,n} = \frac{2}{\sigma_0^2}$. Using (A.10) and the expressions for $\mathbf{B}_{n',n'}$ and $\mathbf{B}_{n,n}$, (A.4) becomes the following:

$$p^{n'}(\mathbf{Y}|\mathbf{x}^1, \ldots, \mathbf{x}^{n'}, \ldots, \mathbf{x}^N) = \frac{1}{(4\pi)^N |K|^{1/2}} \exp(-\frac{y^{n'}}{\sigma_1^2}) \prod_{\substack{n=1 \\ n\neq n'}}^{N} \exp(-\frac{y^n}{\sigma_0^2}). \tag{B.11}$$

with K above a diagonal matrix as we demonstrated above. We emphasize that in the above expression, the assumption is that L targets are present at a locations defined by $\mathbf{x}^{n'}$ proposed by target n'; and is not present in any location $\mathbf{x}^n$ proposed by the rest of the particles $n = 1, \ldots, N, n \neq n'$. The likelihood ratio used in the particle filter weight equation is the ratio of the likelihood when targets are present at $\mathbf{x}^{n'}$ to the likelihood when targets are absent. This is given by the simple expression:

$$\frac{p^{n'}(\mathbf{Y}|\mathbf{x}^1, \ldots, \mathbf{x}^{n'}, \ldots, \mathbf{x}^N)}{p_0^{n'}(\mathbf{Y}|\mathbf{x}^1, \ldots, \mathbf{x}^{n'}, \ldots, \mathbf{x}^N)} = \frac{\exp(-\frac{\mathbf{y}^{n'}}{\sigma_1^2})}{\exp(-\frac{\mathbf{y}^{n'}}{\sigma_0^2})}.$$

Therefore, the covariance of measurements can be controlled by the selection of the waveform transmitted. We conclude, that for waveforms based on Björck CAZAC sequences that have concentrated AFs the measurement independence approximation is reasonable.

APPENDIX C

Sampling Importance Resampling Particle Filter Sample Code

```
% SIR_Particle_Filter.m
% This is a Sample Importance Resampling particle
filtering code
% for estimating the position and velocity of a single target in the
% Cartesian coordinates using radar measurements from two
sensors.

clear
clc

% settings
N=1000; % number of particles to use
endK=100; % number of time steps in the scenario
% kinematic prior standard deviation in position and velocity
sigpos=[5 5]'; % x-y position
sigvel=[2 2]'; % x-y velocity
% time difference between time steps
dt=1;
% measurement noise variance
R=[10 2 10 2]; % corresponding to range and range-rate for sensor
1 and 2 respectively
% the initial true target state
X(1:2,1)=[1000 1000]; % x-y position
v(1:2,1)=[5 5]; % x-y velocity
% generate the true target state throughout the scenario
for tstep=2:endK
X(1:2,tstep) = X(1:2,tstep-1) + v(1:2,tstep-1)*dt +
sigpos(1:2).*randn(2,1);
```

```
v(1:2,tstep) = v(1:2,tstep-1) + sigvel(1:2).*randn(2,1);
end
% true state of the target
v(1,1:endK-1)=(X(1,2:endK)-X(1,1:endK-1))*dt; %x
v(2,1:endK-1)=(X(2,2:endK)-X(2,1:endK-1))*dt; %y
% locations of the two sensors
% sensor 1
Xs(1,1:endK)=2000;
Xs(2,1:endK)=-2000;
%sensor 2
Xs2(1,1:endK)=4000;
Xs2(2,1:endK)=-2000;
% % % uncomment to plot the target true state and sensor
positions
% % figure(1)
% % plot(X(1,1:endK),X(2,1:endK)); hold on
% % plot(Xs(1,1),Xs(2,1),'*')
% % plot(Xs2(1,1),Xs2(2,1),'*'); hold off
% true range and range rate
r(1:endK)=sqrt((Xs(1,1:endK)-X(1,1:endK)).^2+(Xs(2,1:endK)-
X(2,1:endK)).^2); %range, sensor 1
rr(1:endK-1)=(v(1,1:endK-1).*(X(1,1:endK-1)-Xs(1,1:endK-1))
+v(2,1:endK-1).*(X(2,1:endK-1)-Xs(2,1:endK-1)))./r(1:endK-1);
%range rate, sensor 1
r2(1:endK)=sqrt((Xs2(1,1:endK)-X(1,1:endK)).^2+
(Xs2(2,1:endK)-X(2,1:endK)).^2); %range, sensor 2
rr2(1:endK-1)=(v(1,1:endK-1).*(X(1,1:endK-1)-Xs2(1,1:endK-1))
+v(2,1:endK-1).*(X(2,1:endK-1)-Xs2(2,1:endK-1)))./r2(1:endK-
1); %range rate, sensor 2
% add sensor noise to the true range and range-rates
r(1:endK)=r(1:endK) + sqrt(R(1))*randn;
rr(1:endK-1)=rr(1:endK-1) + sqrt(R(2))*randn;
r2(1:endK)=r2(1:endK) + sqrt(R(3))*randn;
rr2(1:endK-1)=rr2(1:endK-1) + sqrt(R(4))*randn;
% initialize particles using true positions
Xpart(1:2,1:N)=X(1:2,1)*ones(1,N) +
sigpos(1:2)*ones(1,N).*randn(2,N); %position
vpart(1:2,1:N)=v(1:2,1)*ones(1,N) +
sigvel(1:2)*ones(1,N).*randn(2,N); %speed
```

```
% use the particles to find the range and range-rates with respect to
each
% sensor
rpart(1:N)=sqrt((Xs(1,1)-Xpart(1,1:N) ).^2+(Xs(2,1)- Xpart(1,1:N)
).^2); % particle range, sensor 1
rrpart(1:N)=(vpart(1,1:N).*(Xpart(1,1:N)-Xs(1,1))
+vpart(2,1:N).*(Xpart(2,1:N)-Xs(2,1)))./rpart(1:N); % particle
range-rate, sensor 1
rpart2(1:N)=sqrt((Xs2(1,1)-Xpart(1,1:N) ).^2+(Xs2(2,1)-
Xpart(1,1:N) ).^2); % particle range, sensor 2
rrpart2(1:N)=(vpart(1,1:N).*(Xpart(1,1:N)-Xs2(1,1))
+vpart(2,1:N).*(Xpart(2,1:N)-Xs2(2,1)))./rpart2(1:N); % particle
range-rate, sensor 2
% determine the initial estimate
Xhat(1:2,1)=Xpart(1:2,:)* ones(N,1)/N;
vhat(1:2,1)=vpart(1:2,:)* ones(N,1)/N;
for tstep=2:endK-1
% Project particles one step ahead using the kinematic prior
Xprev=Xpart;
vprev=vpart;
Xpart(1:2,1:N)=Xprev(1:2,1:N) + vpart(1:2,1:N)*dt +
sigpos(1:2)*ones(1,N).*randn(2,N); % position
vpart(1:2,1:N)=vprev(1:2,1:N) +
sigvel(1:2)*ones(1,N).*randn(2,N); % velocity
% find the range and range rates from the projected particles
with
% respect to the two sensors
rpart(1:N)=sqrt((Xs(1,tstep)-Xpart(1,1:N)).^2+(Xs(2,tstep)-
Xpart(2,1:N)).^2); % range, sensor 1
rrpart(1:N)=(vpart(1,1:N).*(Xpart(1,1:N)-Xs(1,tstep))
+vpart(2,1:N).*(Xpart(2,1:N)-Xs(2,tstep)))./rpart(1:N); % rangerate,
sensor 1
rpart2(1:N)=sqrt((Xs2(1,tstep)-Xpart(1,1:N)).^2+(Xs2(2,tstep)-
Xpart(2,1:N)).^2); % range, sensor 2
rrpart2(1:N)=(vpart(1,1:N).*(Xpart(1,1:N)-Xs2(1,tstep))
+vpart(2,1:N).*(Xpart(2,1:N)-Xs2(2,tstep)))./rpart2(1:N); %
range-rate, sensor 2
% find the weight of each particle using the likelihood
for n=1:N
```

```
% the weight of each particle is proportional to the likelihood
from each sensor
weight(n) = exp(-0.5 * (rpart(n)-r(tstep))^2/R(1)) * ...
exp(-0.5 * (rrpart(n)-rr(tstep))^2/R(2)) * ...
exp(-0.5 * (rpart2(n)-r2(tstep))^2/R(3)) * ...
exp(-0.5 * (rrpart2(n)-rr2(tstep))^2/R(4));
end
if sum(weight)==0
% to avoid having all zero weights
disp('all zero')
Weight = ones(1,N)/N;
else
% normalize weights
Weight = weight/sum(weight);
end
% determine estimate
Xhat(1:2,tstep)=Xpart(1:2,:)*(Weight(:));
vhat(1:2,tstep)=vpart(1:2,:)*(Weight(:));
% % % uncomment to plot particles before resampling
% % figure(2)
% %
plot(X(1,:),X(2,:),'b.',Xhat(1,tstep),Xhat(2,tstep),'r*',Xpart(1,1:N),
Xpart(2,1:N),'r.'); hold on
% % plot(X(1,tstep),X(2,tstep),'g.'); hold off
% % pause(.1)
% resample partcles
% use the cumulative distribution function
cumpr = cumsum(Weight(1,1:length(Weight)))';
cumpr = cumpr/max(cumpr);
u(1,1) = (1/N)*rand(1,1);
i=1;
for j = 1:N
u(j,1)= u(1,1) + (1/N)*(j-1);
while (u(j,1) > cumpr(i,1))
i = i+1;
if i > N
break
end
end
```

```
if i <= N
inew(j) = i;
end
if (j<=N) & (i>N)
inew(j:N) = N* ones(1,N-j+1);
end
end
Xpart(:,1:N)=Xpart(:,inew);
vpart(:,1:N)=vpart(:,inew);
% % % uncomment to plot particles after resampling
% % figure(2)
% %
plot(X(1,:),X(2,:),'b.',Xhat(1,tstep),Xhat(2,tstep),'r*',Xpart(1,1:N),
Xpart(2,1:N),'r.'); hold on
% % plot(X(1,tstep),X(2,tstep),'g.'); hold off
% % pause(.1)
end %tstep
figure(3)
plot(X(1,:),X(2,:),'b.',Xhat(1,:),Xhat(2,:),'r*'); hold on
plot(Xs(1,1),Xs(2,1),'*')
plot(Xs2(1,1),Xs2(2,1),'*'); hold off
```

APPENDIX D

Likelihood Particle Filter Sample Code

```
% Likelihood_Particle_Filter.m
% This is a Likelihood Particle
Filtering code
% for estimating the position and velocity of a single target in the
% Cartesian coordinates using radar measurements from two
sensors.

clear
clc

% settings
N=100; % number of particles to use
endK=100; % number of time steps in the scenario
% kinematic prior standard deviation in position and velocity
sigpos=[5 5]'; % x-y position
sigvel=[2 2]'; % x-y velocity
% time difference between time steps
dt=1;
% measurement noise variance
% (note that the likelihood is highly peaked in this case. Therefore
a likelihood particle filter is needed)
R=[1 .2 1 .2]; % corresponding to range and range-rate for sensor
1 and 2 respectively
% the initial true target state
X(1:2,1)=[1000 1000]; % x-y position
v(1:2,1)=[5 5]; % x-y velocity
% generate the true target state throughout the scenario
for tstep=2:endK
X(1:2,tstep) = X(1:2,tstep-1) + v(1:2,tstep-1)*dt +
sigpos(1:2).*randn(2,1);
```

```
v(1:2,tstep) = v(1:2,tstep-1) + sigvel(1:2).*randn(2,1);
end
% true state of the target
v(1,1:endK-1)=(X(1,2:endK)-X(1,1:endK-1))*dt; %x
v(2,1:endK-1)=(X(2,2:endK)-X(2,1:endK-1))*dt; %y
% locations of the two sensors
% sensor 1
Xs(1,1:endK)=2000;
Xs(2,1:endK)=-2000;
% sensor 2
Xs2(1,1:endK)=4000;
Xs2(2,1:endK)=-2000;
% % % uncomment to plot the target true state and sensor
positions
% % figure(1)
% % plot(X(1,1:endK),X(2,1:endK)); hold on
% % plot(Xs(1,1),Xs(2,1),'*')
% % plot(Xs2(1,1),Xs2(2,1),'*'); hold off
% true range and range rate
r(1:endK)=sqrt((Xs(1,1:endK)-X(1,1:endK)).^2+(Xs(2,1:endK)-
X(2,1:endK)).^2); %range, sensor 1
rr(1:endK-1)=(v(1,1:endK-1).*(X(1,1:endK-1)-Xs(1,1:endK-1))
+v(2,1:endK-1).*(X(2,1:endK-1)-Xs(2,1:endK-1)))./r(1:endK-1);
%range rate, sensor 1
r2(1:endK)=sqrt((Xs2(1,1:endK)-X(1,1:endK)).^2+
(Xs2(2,1:endK)-X(2,1:endK)).^2); %range, sensor 2
rr2(1:endK-1)=(v(1,1:endK-1).*(X(1,1:endK-1)-Xs2(1,1:endK-1))
+v(2,1:endK-1).*(X(2,1:endK-1)-Xs2(2,1:endK-1)))./r2(1:endK-
1); %range rate, sensor 2
% add sensor noise to the true range and range-rates
r_noisy(1:endK)=r(1:endK) + sqrt(R(1))*randn;
rr_noisy(1:endK-1)=rr(1:endK-1) + sqrt(R(2))*randn;
r2_noisy(1:endK)=r2(1:endK) + sqrt(R(3))*randn;
rr2_noisy(1:endK-1)=rr2(1:endK-1) + sqrt(R(4))*randn;
% initialize particles using true positions
Xpart(1:2,1:N)=X(1:2,1)*ones(1,N) +
sigpos(1:2)*ones(1,N).*randn(2,N); %position
vpart(1:2,1:N)=v(1:2,1)*ones(1,N) +
sigvel(1:2)*ones(1,N).*randn(2,N); %speed
```

```
% use the particles to find the range and range-rates with respect to
each
% sensor
rpart(1:N)=sqrt((Xs(1,1)-Xpart(1,1:N) ).^2+(Xs(2,1)- Xpart(1,1:N)
).^2); % particle range, sensor 1
rrpart(1:N)=(vpart(1,1:N).*(Xpart(1,1:N)-Xs(1,1))
+vpart(2,1:N).*(Xpart(2,1:N)-Xs(2,1)))./rpart(1:N); % particle
range-rate, sensor 1
rpart2(1:N)=sqrt((Xs2(1,1)-Xpart(1,1:N) ).^2+(Xs2(2,1)-
Xpart(1,1:N) ).^2); % particle range, sensor 2
rrpart2(1:N)=(vpart(1,1:N).*(Xpart(1,1:N)-Xs2(1,1))
+vpart(2,1:N).*(Xpart(2,1:N)-Xs2(2,1)))./rpart2(1:N); % particle
range-rate, sensor 2
rrpartnonoise(1:N)=rrpart(1:N);
rrpartnonoise2(1:N)=rrpart2(1:N);
% determine the initial estimate
Xhat(1:2,1)=Xpart(1:2,:)* ones(N,1)/N;
vhat(1:2,1)=vpart(1:2,:)* ones(N,1)/N;
for tstep=2:endK-1
Xprev=Xpart;
vprev=vpart;
% project the particles ahead
Xpr(1:2,1:N) = Xprev(1:2,1:N) + vpart(1:2,1:N)*dt;
rpartnonoise(1:N)=sqrt((Xs(1,tstep)-Xpr(1,1:N) ).^2+(Xs(2,tstep)-
Xpr(2,1:N) ).^2); % projected range, sensor 1
rpartnonoise2(1:N)=sqrt((Xs2(1,tstep)-Xpr(1,1:N) ).^2+
(Xs2(2,tstep)- Xpr(2,1:N) ).^2); % projected range, sensor 2
% find the range of the range/range-rate resolution cells according
to the kinematic prior to
% sample with the likelihood
rpartmin(1:N)=rpartnonoise(1:N)-norm(3*sigpos(1:2));
rpartmax(1:N)=rpartnonoise(1:N)+norm(3*sigpos(1:2));
rpartmin2(1:N)=rpartnonoise2(1:N)-norm(3*sigpos(1:2));
rpartmax2(1:N)=rpartnonoise2(1:N)+norm(3*sigpos(1:2));
rrpartmin(1:N)=rrpartnonoise(1:N)-sqrt(2)*3*sigvel(1);
rrpartmax(1:N)=rrpartnonoise(1:N)+sqrt(2)*3*sigvel(1);
rrpartmin2(1:N)=rrpartnonoise2(1:N)-sqrt(2)*3*sigvel(1);
rrpartmax2(1:N)=rrpartnonoise2(1:N)+sqrt(2)*3*sigvel(1);
for n=1:N
```

```
% for each particle find the proposed range/range-rates
res=1000; % set the resolution
rpartvect(:,n)=rpartmin(n):(rpartmax(n)-
rpartmin(n))/res:rpartmax(n);
rpartvect2(:,n)=rpartmin2(n):(rpartmax2(n)-
rpartmin2(n))/res:rpartmax2(n);
rrpartvect(:,n)=rrpartmin(n):(rrpartmax(n)-
rrpartmin(n))/res:rrpartmax(n);
rrpartvect2(:,n)=rrpartmin2(n):(rrpartmax2(n)-
rrpartmin2(n))/res:rrpartmax2(n);
% weight each of the proposed range/range-rates for each sensor
for indcell=1:length(rpartvect(:,n))
% the weight of each proposed cell is proportional to the
likelihood from each sensor
weightcellr(indcell,n) = exp(-0.5 * (rpartvect(indcell,n)-
r_noisy(tstep))^2/R(1));
weightcellrr(indcell,n) = exp(-0.5 * (rrpartvect(indcell,n)-
rr_noisy(tstep))^2/R(2));
end % indcell
for indcell=1:length(rpartvect(:,n))
% the weight of each proposed cell is proportional to the
likelihood from each sensor
weightcellr2(indcell,n) = exp(-0.5 * (rpartvect2(indcell,n)-
r2_noisy(tstep))^2/R(1));
weightcellrr2(indcell,n) = exp(-0.5 * (rrpartvect2(indcell,n)-
rr2_noisy(tstep))^2/R(2));
end % indcell
% sensor 1 range
if sum(weightcellr(:,n))==0
disp('all zero')
% to avoid having all zero weights
Weightcellr(:,n) = 1/length(rpartvect(:,n));
Normfctr(n)=1;
else
% normalize weights
Normfctr(n) = sum(weightcellr(:,n));
Weightcellr(1:length(weightcellr(:,n)),n) =
weightcellr(:,n)/sum(weightcellr(:,n));
end
```

```
% sensor 1 range rate
if sum(weightcellrr(:,n))==0
disp('all zero')
% to avoid having all zero weights
Weightcellrr(:,n) = 1/length(rpartvect(:,n));
Normfctrr(n)=1;
else
% normalize weights
Normfctrr(n) = sum(weightcellrr(:,n));
Weightcellrr(1:length(weightcellrr(:,n)),n) =
weightcellrr(:,n)/sum(weightcellrr(:,n));
end
% sensor 2 range
if sum(weightcellr2(:,n))==0
disp('all zero')
% to avoid having all zero weights
Weightcellr2(:,n) = 1/length(rpartvect(:,n));
Normfctr2(n)=1;
else
% normalize weights
Normfctr2(n) = sum(weightcellr2(:,n));
Weightcellr2(1:length(weightcellr2(:,n)),n) =
weightcellr2(:,n)/sum(weightcellr2(:,n));
end
% sensor 2 range rate
if sum(weightcellrr2(:,n))==0
disp('all zero')
% to avoid having all zero weights
Weightcellrr2(:,n) = 1/length(rpartvect(:,n));
Normfctrr2(n)=1;
else
% normalize weights
Normfctrr2(n) = sum(weightcellrr2(:,n));
Weightcellrr2(1:length(weightcellrr2(:,n)),n) =
weightcellrr2(:,n)/sum(weightcellrr2(:,n));
end
% this code samples 1 sample given the weight distribution
% sensor 1
cumprcumpr = cumsum(Weightcellr(:,n)); % cumulative
```

```
distribution function (uniform)
cumprcumpr = cumprcumpr/max(cumprcumpr);
res1=cumprcumpr-rand;
inew1r=find(res1 >=0,1);
cumprcumpr = cumsum(Weightcellrr(:,n)); % cumulative
distribution function (uniform)
cumprcumpr = cumprcumpr/max(cumprcumpr);
res1=cumprcumpr-rand;
inew1rr=find(res1 >=0,1);
% sensor 2
cumprcumpr = cumsum(Weightcellr2(:,n)); % cumulative
distribution function (uniform)
cumprcumpr = cumprcumpr/max(cumprcumpr);
res1=cumprcumpr-rand;
inew2r=find(res1 >=0,1);
cumprcumpr = cumsum(Weightcellrr2(:,n)); % cumulative
distribution function (uniform)
cumprcumpr = cumprcumpr/max(cumprcumpr);
res1=cumprcumpr-rand;
inew2rr=find(res1 >=0,1);
% set the cell that was sampled based on the likelihood to be the
one
% proposed by each particle
rpart(n)=rpartvect(inew1r,n);
rrpart(n)=rrpartvect(inew1rr,n);
rpart2(n)=rpartvect2(inew2r,n);
rrpart2(n)=rrpartvect2(inew2rr,n);
% with the sampled rate and rate-rates, we find the state in the
Cartesian coordinates
% this finds the intersection of the two circles defined by the two
ranges and the sensor locations
[i1, i2] = circleintersect2(Xs(:,tstep)', rpart(n), Xs2(:,tstep)',
rpart2(n)); % 'circleintersect' code from
http://www.csse.uwa.edu.au/~pk/research/matlabfns/Projective/cir
cleintersect.m
% out of these two solutions we find the one that agrees with the
kinematics. Hopefully, the solution closer to the propagated
noiseless particle location is
% the correct
```

```
if norm(Xpr(1:2,n)-i1')<norm(Xpr(1:2,n)-i2')
Xpart(:,n)=i1; %select i1
else
Xpart(:,n)=i2;
end
end %n
vpart(1,1:N) = (rrpart2(1:N).*rpart2(1:N).*(Xpart(2,1:N)-Xs(2)) -
rrpart(1:N).*rpart(1:N).*(Xpart(2,1:N)-Xs2(2)) ) ...
/ ( (Xpart(1,1:N)-Xs2(1)).*(Xpart(2,1:N)-Xs(2)) -
(Xpart(1,1:N)-Xs(1)).*(Xpart(2,1:N)-Xs2(2)) );
vpart(2,1:N) = ( rrpart(1:N).*rpart(1:N) - (Xpart(1,1:N)-
Xs(1)).*vpart(1,1:N) ) / ( Xpart(2,1:N)-Xs(2) ) ;
for n=1:N
% find the weight of each particle using the likelihood
% the weight of each particle is proportional to the likelihood
from each sensor
weight(n) = Normfctr(n) * Normfctrr(n) * Normfctr2(n) *
Normfctrr2(n) *...
exp(-0.5 * (rpart(n)-r(tstep))^2/R(1)) * ...
exp(-0.5 * (rrpart(n)-rr(tstep))^2/R(2)) * ...
exp(-0.5 * (rpart2(n)-r2(tstep))^2/R(3)) * ...
exp(-0.5 * (rrpart2(n)-rr2(tstep))^2/R(4));
end %n
if sum(weight)==0
% to avoid having all zero weights
disp('all zero')
Weight = ones(1,N)/N;
else
% normalize weights
Weight = weight/sum(weight);
end
% determine estimate
Xhat(1:2,tstep)=Xpart(1:2,:)*(Weight(:));
vhat(1:2,tstep)=vpart(1:2,:)*(Weight(:));
% % % uncomment to plot particles before resampling
% % figure(2)
% %
plot(X(1,:),X(2,:),'b.',Xhat(1,tstep),Xhat(2,tstep),'r*',Xpart(1,1:N),
Xpart(2,1:N),'r.'); hold on
```

```
% % plot(X(1,tstep),X(2,tstep),'g.'); hold off
% % pause(.1)
% resample partcles
% use the cumulative distribution function
cumpr = cumsum(Weight(1,1:length(Weight)))';
cumpr = cumpr/max(cumpr);
u(1,1) = (1/N)*rand(1,1);
i=1;
for j = 1:N
u(j,1)= u(1,1) + (1/N)*(j-1);
while (u(j,1) > cumpr(i,1))
i = i+1;
if i > N
break
end
end
if i <= N
inew(j) = i;
end
if (j<=N) & (i>N)
inew(j:N) = N* ones(1,N-j+1);
end
end
Xpart(:,1:N)=Xpart(:,inew);
% sampled range-rate carried to the next time step
rrpartnonoise(1:N)=rrpart(inew);
rrpartnonoise2(1:N)=rrpart2(inew);
% % % uncomment to plot particles after resampling
% % figure(2)
% %
plot(X(1,:),X(2,:),'b.',Xhat(1,tstep),Xhat(2,tstep),'r*',Xpart(1,1:N),
Xpart(2,1:N),'r.'); hold on
% % plot(X(1,tstep),X(2,tstep),'g.'); hold off
% % pause(.01)
end %tstep
figure(3)
plot(X(1,:),X(2,:),'b.',Xhat(1,:),Xhat(2,:),'r*'); hold on
plot(Xs(1,1),Xs(2,1),'*')
plot(Xs2(1,1),Xs2(2,1),'*'); hold off
```

Bibliography

[1] N. Levanon and E. Mozeson, *Radar Signals,* Wiley, 2004. DOI: 10.1002/0471663085 1, 2, 8, 25, 65

[2] C. Rago, P. Willett, Y. Bar-Shalom, "Detection-Tracking Performance With Combined Waveforms," *IEEE Trans. on Aerospace and Electronic Systems,* vol. 34, no. 2, pp. 612–624, April 1998. DOI: 10.1109/7.670395 1, 2, 26, 27, 33, 42, 69

[3] Ruixin Niu, P. Willett, Y. Bar-Shalom, "Tracking Considerations in Selection of Radar Waveform for Range and Range-Rate Measurements," *IEEE Trans. on Aerospace and Electronic Systems,* vol. 38, no. 2, pp. 467–487, April, 2002. DOI: 10.1109/TAES.2002.1008980 1, 2, 27

[4] M. S. Arulampalam, S. Maskell, N. Gordon, and T. Clapp "A Tutorial on Particle Filters for Online Nonlinear/Non-Gaussian Bayesian Tracking", *IEEE Transactions on Signal Processing,* vol. 50, no. 2, pp. 174–188, Feb. 2002. DOI: 10.1109/78.978374 2, 17, 18, 21, 22, 31, 32, 34, 37, 46, 53

[5] I. Kyriakides, I. Konstantinidis, D. Morrell, J. J. Benedetto, and A. Papandreou-Suppappola, "Target Tracking Using Particle Filtering and CAZAC Sequences," *Waveform Design and Diversity Conference,* pp. 367–371, June 2007. DOI: 10.1109/WDDC.2007.4339445 2, 47, 66

[6] G. Björck, "Functions of Modulus One on $\mathbb{Z}_n$ Whose Fourier Transforms have Constant Modulus, and Cyclic n-Roots," *Proc. of 1989 NATO Advanced Study Institute on Recent Advances in Fourier Analysis and Its Applications,* J.S. Byrnes and J. L. Byrnes, ed. pp. 131–140, 1990. 2, 5

[7] J.J. Benedetto, J. Donatelli, I. Konstantinidis and C. Shaw, "A Doppler Statistic for Zero Autocorrelation Waveforms," *40th Annual Conference on Information Sciences and Systems,* pp. 1403–1407, March 2006. DOI: 10.1109/CISS.2006.286684 2, 65

[8] I. Kyriakides, T. Trueblood, D. Morrell, and A. Papandreou-Suppappola, "Multiple Target Tracking Using Likelihood Particle Filtering and Adaptive Waveform Design," Asilomar Conference on Signals, Systems, and Computers, pp. 1188–1192, Oct. 2008. DOI: 10.1109/ACSSC.2008.5074603 2

[9] I. Kyriakides, D. Morrell, and A. Papandreou-Suppappola, "Multiple Target Tracking Using Particle Filtering and Adaptive Multicarrier Phase-Coded CAZAC Sequences," IEEE Transactions on Signal Processing, (to be submitted). 2

[10] D. J. Kershaw and R. J. Evans, "Optimal Waveform Selection for Tracking Systems," *IEEE Trans. Inform. Theory*, vol. 40, no. 5, pp. 1536–1550, Sep. 1994. DOI: 10.1109/18.333866 2

[11] D. J. Kershaw and R. J. Evans, "Waveform Selective Probabilistic Data Association," *IEEE Trans. Aerosp. Electron. Syst.*, vol. 33, pp. 1180–1188, Oct. 1997. DOI: 10.1109/7.625110 2

[12] S.P. Sira, A. Papandreou-Suppappola and D. Morrell, "Dynamic Configuration of Time-Varying Waveforms for Agile Sensing and Tracking in Clutter," *IEEE Transactions on Signal Processing*, vol. 55, no. 7, pp. 3207–3217, July 2007. DOI: 10.1109/TSP.2007.894418 2, 42

[13] S. P. Sira, A. Papandreou-Suppappola, and D. Morrell, "Time-Varying Waveform Selection and Configuration for Agile Sensors in Tracking Applications," *IEEE International Conference on Acoustics, Speech, and Signal Processing*, vol. 5, pp. 881–884, Mar. 2005. DOI: 10.1109/ICASSP.2005.1416445 2

[14] J.J. Kroszczynski, "Pulse Compression by Means of Linear-Period Modulation," *Proceedings of the IEEE* vol. 57, no. 7, pp. 1260–1266, July 1969. DOI: 10.1109/PROC.1969.7230 5

[15] J.J. Benedetto, J.J. Donatelli, "Ambiguity Function and Frame-Theoretic Properties of Periodic Zero-Autocorrelation Waveforms," *IEEE Journal of Selected Topics in Signal Processing*, vol. 1, no. 1, pp. 6–20, June 2007. DOI: 10.1109/JSTSP.2007.897044 5, 7

[16] J. Benedetto, A. Bourouihiya, I. Konstantinidis, K. Okoudjou, "Concatenating codes for improved ambiguity behavior," *Int. Conf. on Electromagnetics in Advanced Applications*, Sept. 2007. DOI: 10.1109/ICEAA.2007.4387337 5, 6

[17] A. Kebo, I. Konstantinidis, J. J. Benedetto, M. R. Dellomo, J. M. Sieracki, "Ambiguity and sidelobe behavior of CAZAC coded waveforms," *IEEE Radar Conference*, pp. 99–103, Boston, MA, 2007. DOI: 10.1109/RADAR.2007.374198 6

[18] H. L. Van Trees, *Detection Estimation and Modulation Theory, Part III.* New York: Wiley, 1971. 6

[19] S.S. Blackman, "Multiple Target Tracking With Radar Applications," Archtech House, Norwood, MA, 1986. 17

[20] M. Montemerlo, S. Thrun, and W. Whittacker, "Conditional Particle Filter for Simultaneous Mobile Robot Localization and People Tracking," *Proceedings of the IEEE Conference on Robotics and Automation*, vol. 1, pp. 695–701, 2002. DOI: 10.1109/ROBOT.2002.1013439 17

[21] D. Tweed and A. Calway, "Tracking Multiple Animals in Wildlife Footage," *Proceedings of the Conference on Pattern Recognition*, vol. 2, pp. 24–27, 2002. DOI: 10.1109/ICPR.2002.1048227 17

[22] D. Schulz, W. Burgard, D. Fox and A. B. Cremers, "Tracking Multiple Moving Targets with a Mobile Robot using Particle Filters and Statistical Data Association," *Proceedings of the IEEE International Conference on Robotics and Automation*, 2001. DOI: 10.1109/ROBOT.2001.932850 17

[23] M. Isard and J. MacCormic, "BraMBLe: A Bayesian Multiple-Blob Tracker," *Proceedings of the 8th International Conference on Computer Vision*, 2001. DOI: 10.1109/ICCV.2001.937594 17

[24] R. E. Kalman, "A New Approach to Linear Filtering and Prediction Problems," *Transactions of the ASME–Journal of Basic Engineering*, vol. 82, series D, pp. 35–45, 1960. 17

[25] C. R. Rao, C. R. Sastry and Y. Zhou, "Tracking the Direction-of-Arrival of Multiple Moving Targets," *IEEE Transactions on Signal Processing*, vol. 42, Issue 5, pp. 1133–1144, 1994. DOI: 10.1109/78.295205 17

[26] A. H. Jazwinsky, "Stochastic Processes and Filtering Theory," New York: Academic Press, 1970. 17

[27] A. J. Julier, and J. K. Uhlman, "A New Extension of the Kalman Filter to Nonlinear Systems," *Proceedings of Aerosence: The Eleventh International Symposium on Aerospace/Defence Sensing, Simulation and Controls*, vol. 3068, pp. 182–193, 1997. 17

[28] L. D. Stone, C. A. Barlow, and T. L. Corwin, *Bayesian Multiple Target Tracking*, Artech House, 1999. 18

[29] A. Doucet, S. Godsill, and C. Andreu, "On Sequential Monte Carlo Sampling Methods for Bayesian Filtering," *Statistics and Computing*, vol. 10, no. 3, pp. 197–208, 2000. DOI: 10.1023/A:1008935410038 19, 20, 21

[30] A. Doucet, J. F. G. de Freitas, and N. J. Gordon, "An Introduction to Sequential Monte Carlo Methods," *Sequential Monte Carlo Methods in Practice*, A. Doucet, J. F. G. de Freitas, and N. J. Gordon, Eds. New York: Springer-Verlag, 2001. 19

[31] J. Carpender, P. Clifford, and P. Fearnhead, "Improved Particle Filter for Nonlinear Problems," *Proceedings of the Institute of Electrical Engineering, Radar, Sonar, Navigation*, 1999. DOI: 10.1049/ip-rsn:19990255 19

[32] N. Gordon, D. Salmond, and A. F. M. Smith, "Novel Approach to Nonlinear, and Non-Gaussian Bayesian State Estimation," *Proceedings of the Institute of Electrical Engineering*, vol. 140, pp. 107–113, 1993. 19

[33] J. MacCormick and A. Blake, "A Probabilistic Exclusion Principle for Tracking Multiple Objects," *Proceedings of the International Conference of Computer Vision*, 1999, pp. 572–578. DOI: 10.1109/ICCV.1999.791275 19

[34] D. Crisan, P. Del Moral, and T. J. Lyons, "Nonlinear Filtering Using Branching and Interacting Particle Systems," *Markov Processes Related Fields*, vol. 5, no. 3, pp. 293–319, 1999. 19

[35] P. Del Moral, "Nonlinear Filtering: Interacting Particle Solution," *Markov Processes Realted Fields*, vol. 2, no. 4, pp. 555–580. 19

[36] K. Kanazawa, D. Koller, and S. J. Russell, "Stochastic Simulation Algorithms for Dynamic Probabilistic Networks," *Proceedings of the Eleventh Annual Conference of Uncertainty AI*, 1995, pp. 346–351. 19

[37] N. Bergman, "Recursive Bayesian Estimation: Navigation and Tracking Applications," Ph.D. dissertation, Linköping Univ., Linköping, Sweden, 1999. 20

[38] D. Crisan and A. Doucet, "A Survey of Theoretical Results on Particle Filtering for Practitioners," *IEEE Trans. Signal Processing*, vol. 50, no. 4, pp. 736–746, 2002. DOI: 10.1109/78.984773 21

[39] G. Kitawa, "Monte Carlo Filter and Smoother for Non-Gaussian Nonlinear State Space Models," *Journal of Computational and Graphical Statistics*, vol. 5(1), pp. 1–25, 1996. DOI: 10.2307/1390750 21

[40] J. S. Liu and R. Chen, "Sequential Monte Carlo Methods for Dynamical Systems," *Journal of American Statistical Association*, vol. 93, pp. 1032–1044, 1998. DOI: 10.2307/2669847 21

[41] C. Kreucher, K. Kastella and A. O. Hero III, "Tracking Multiple Targets Using a Particle Filter Representation of the Joint Multitarget Probability Density", *SPIE Int. Symp. on Optical Science and Techn.*, vol. 5204, pp. 258–259, 2003. DOI: 10.1117/12.502696 22, 46

[42] M. Orton and W. Fitzgerald, "A Bayesian Approach to Tracking Multiple Targets Using Sensor Arrays and Particle Filters," *IEEE Transactions on Signal Processing*, vol. 50, no. 2, pp. 216–223, Feb. 2002. DOI: 10.1109/78.978377 22, 25, 41, 46

[43] P. L. Bogler, *Radar Principles With Applications to Tracking Systems,* New York, John Wiley & Sons, 1990. 25

[44] M. I. Skolnik, *Introduction to Radar Systems,* McGraw-Hill, 1980. 26, 42, 65, 75

[45] R.K. Mallik, "On Multivariate Rayleigh and Exponential Distributions," *IEEE Transactions on Information Theory*, vol. 49, no. 6, pp. 1499–1515, June 2003. DOI: 10.1109/TIT.2003.811910 32, 51, 53, 66, 76

[46] G.J. Foster, J.J. Petruzzo III, T.N. Phan, "Track filtering of boosting targets," *Proceedings of the 35th Southeastern Symposium on System Theory*, vol. 35, pp. 450–454, March 2003. DOI: 10.1109/SSST.2003.1194611 41

[47] E. F. Knott, J. F. Shaeffer, and M. T. Tuley, *Radar Cross Section,* 2nd Edition, SciTech Publishing, 2004. 42

[48] I. Kyriakides, D. Morrell and A. Papandreou-Suppappola, "Sequential Monte Carlo Methods for Tracking Multiple Targets with Deterministic and Stochastic Constraints," *IEEE Transactions on Signal Processing,* vol. 56, no. 3, pp. 937–948, March 2008. DOI: 10.1109/TSP.2007.908931 46

Authors' Biographies

IOANNIS KYRIAKIDES

Ioannis Kyriakides received his B.S. degree in Electrical Engineering in 2003 from Texas A&M University. He received his M.S. and Ph.D. degrees in 2005 and 2008, respectively, from Arizona State University. His research interests include Bayesian target tracking, sequential Monte Carlo methods, radar waveform design, and compressive sensing and processing. He is currently a lecturer at the Electrical and Computer Engineering Department of the University of Nicosia.

DARRYL MORRELL

Darryl Morrell received his B.S., M.S., and Ph.D. degrees in Electrical Engineering in 1984, 1986, and 1988 from Brigham Young University. He is currently an Associate Professor at Arizona State University in the Department of Engineering; as the Associate Chair, he is participating in the implementation of a multi-disciplinary undergraduate engineering program using innovative, research-based pedagogical and curricular approaches. His technical research interests include stochastic decision theory applied to sensor scheduling and information fusion and application of research based pedagogy to engineering education.

ANTONIA PAPANDREOU-SUPPAPPOLA

Antonia Papandreou-Suppappola is a Professor in the School of Electrical, Computer and Energy Engineering at Arizona State University. Her research interests and expertise are in the areas of Adaptive Waveform Design for Agile Sensing, Time-varying Signals and Systems Processing, and Stochastic Processing for Detection, Estimation and Tracking. Her funded research work on sensing and information processing includes the development of optimal waveform selection and configuration algorithms using sequential Monte Carlo and stochastic approximation techniques for the detection and tracking of targets in diverse environments; these include underwater, wideband, or dispersive environments, environments with high noise or clutter, urban terrain or requiring multiple or multi-modal sensors.

Printed in the United States
by Baker & Taylor Publisher Services